T0257713

ASTRONOMY

ASTRONOMY
A Self-Teaching Guide

Eighth Edition

Dinah L. Moché, Ph.D.

WILEY

John Wiley & Sons, Inc.

Astronomy: A Self-Teaching Guide, Eighth Edition

Wiley General Trade, an imprint of Turner Publishing Company
424 Church Street • Suite 2240 • Nashville, Tennessee 37219
445 Park Avenue • 9th Floor • New York, New York 10022
www.turnerpublishing.com

Library of Congress Cataloging-in-Publication Data

Moché, Dinah L., 1936- author.
 Astronomy : a self-teaching guide /Dinah L. Moché Ph.D. -- Eighth edition.
    pages cm
 ISBN 978-0-470-23083-1 (acid-free paper)
 ISBN 978-1-684-42950-9 (hardcover)
 1. Astronomy. I. Title.
 QB 45.2.M63 2014
 520--dc23
                            2014016474

Printed in the United States of America

10 9 8 7 6 5 4 3 2 1

CREDITS

We thank the American Astronomical Society for providing dozens of press releases, fact sheets, and celestial images. This service enables us to keep this text accurate and up to date.

Figures are courtesy of the following organizations and individuals:

David Aguilar/Harvard-Smithsonian Center for Astrophysics: 7.8

California Association for Research in Astronomy (adapted from): 2.15

CSIRO: 2.17, 6.19c

ESA/NASA/SOHO: 4.1, 4.8

ESO: 6.3, 6.16c

Ann Feild/Space Science Telescope Institute: 6.23

Gemini Observatory/Neelon Crawford-Polar Fine Arts: 2.6

Hale Observatories: 6.4, 9.28, 11.5, 11.7

The Hubble Heritage Team (AURA/STScI/NASA): 6.21

Hubble Space Telescope WFPC Team, NASA, STScI: 6.16a

Dr. Thomas Jarrett: 6.17

JAXA: 4.9

Lowell Observatory: 9.22

Dinah L. Moché/George Tremberger Jr.: I.2, I.3, 1.12, 2.4, 3.8, 3.9, 3.16, 3.17, 3.18, 3.20, 4.6, 5.3, 5.10 ([selected] data from Barbara J. Anthony-Twarog), 6.6, 6.7, 6.8, 6.12, 6.24a, 6.24b, 6.25, 7.1, 7.2, 7.4, 8.1 ([selected] from NASA), 8.2, 8.4, 8.11, 8.12, 9.12, 9.14, 9.15, 10.8

NASA: I.1, 5.1a, 5.14, 8.14 (adapted), 8.16, 9.2, 9.6, 9.8, 9.9, 9.10, 9.18, 9.19, 9.21, 9.23, 9.26, 9.27, 10.1, 10.5, 10.6, 10.7, 11.1, 11.4, 12.3, 12.4, 12.5

NASA, Reta Beebe, and Amy Simon (New Mexico State University): 9.20

NASA/CXC/ASU/J. Hester, et al.: 5.12b

NASA/CXC/CfA/R. Kraft, et al.: 6.19a

NASA/CXC/MIT/F. K. Baganoff, et al.: 6.11

NASA and ESA: 6.16d

NASA/ESA/ASU/J. Hester & A. Loll: 5.12a

NASA and STScI: 6.16f

NASA, ESA, S. Beckwith (STScI), and the HUDF Team: I.4

NASA, ESA, R. Gendler, T. Lauer (NOAO/AURA/NSF), and A. Feild (STScI): 6.14

NASA, ESA, and the Hubble Heritage (STScI/AURA)-ESA/Hubble Collaboration: 6.16b, 6.16e

NASA, ESA, the Hubble Heritage (STScI/AURA)-ESA/Hubble Collaboration, and A. Evans (University of Virginia, Charlottesville/NRAO/Stony Brook University): 6.20

NASA and the Hubble Heritage Team (STScI/AURA): 5.9, 6.2

NASA, ESA, and T. Lauer (NOAO/AURA/NSF): 6.9

NASA/Johns Hopkins University Applied Physics Laboratory/Carnegie Institution of Washington: 9.1a, 9.1b

NASA/JPL-Caltech: 6.10, 9.17

NASA/JPL-Caltech/M. Meixner (STScI) & the SAGE Legacy Team: 6.13

NASA/JPL-Caltech/T. Pyle (SSC): 12.2

NASA/JPL-Caltech/Univ. Minn./R. Gehrz : 5.12c

NASA/JPL/Space Science Institute: 9.24

NASA/JPL/STScI: 9.25

NASA, Steve Lee (University of Colorado), Jim Bell (Cornell University), and Mike Wolff (Space Science Institute): 9.16

NASA/WMAP Science Team: 7.9

National Optical Astronomy Observatories: 1.3, 5.11, 6.1, 6.5, 6.22a

National Radio Astronomy Observatory/AUI, J. O. Burns, E. J. Schrier, and E. D. Feigelson: 6.18

Image courtesy of NRAO/AUI and Earth image courtesy of the SeaWiFS Project NASA/ GSFC and ORBIMAGE: 2.18

J. William Schopf, Elso S. Barghoorn, Morton D. Masser, and Robert O. Gordon: 12.1

Dr. Martin Schwartzchild, Princeton University: 4.10

Seth Shostak: 12.7

Aurore Simonnet, Sonoma State University, NASA E/PO: 2.19

SOHO (ESA & NASA): 4.7a, 4.12, 4.13

SOHO (ESA & NASA); assembled by Steele Hill (NASA GSFC): 4.11, 4.12, 4.13

SOHO/MDI magnetic map, white light, TRACE 1700A continuum, TRACE Lyman alpha, TRACE 171Å, TRACE 195Å, TRACE 284Å, YOHKOH/SXT X-ray image; composite prepared by Joe Covington (Lockheed Martin Missiles and Space, Palo Alto): 4.7b (clockwise from top)

STScI and NASA: 2.12

Tass/Sovfoto: 9.7

United States Air Force: 11.10

Courtesy John Walker: 4.5

Ryan Wyatt (adapted from): 8.3

P. Marenfeld & NOAO/AURA/NSF: 1.3

Photo Insert

Page 1 top: NASA/Lockheed Martin [images from the NASA Transition Region and Coronal Explorer (TRACE), the Extreme ultraviolet Imaging Telescope (EIT), the Large Angle and Spectrometric coronagraph (LASCO), and the Michelson Doppler Imager (MDI) telescopes on the ESA/NASA Solar and Heliospheric Observatory (SOHO)]; page 1 bottom: Courtesy NASA/JPL-Caltech; page 2 top: X-ray: NASA/CXC/CfA/R. Kraft et al; Radio: NSF/VLA/Univ. Hertfordshire/M. Hardcastle; Optical: ESO/VLT/ISAAC/M. Rejkuba, et al.; page 2 bottom: NASA, ESA, and the Hubble Heritage Team (STScI/AURA); page 3 top left and right: NASA, ESA, and the Hubble Heritage Team (STScI/AURA); page 3 bottom: NASA/WMAP Science Team; page 4 top: X-ray: NASA/CXC/Wesleyan Univ./R. Kilgard, et al.; UV: NASA/JPL-Caltech; Optical: NASA/ESA/S. Beckwith & Hubble Heritage Team (STScI/AURA); IR: NASA/JPL-Caltech/ Univ. of AZ/R. Kennicutt; page 4 bottom: X-ray: NASA/UMass/Q. D. Wang, et al.; Optical: NASA/STScI/AURA/Hubble Heritage; Infrared: NASA/JPL-Caltech/Univ. AZ/R. Kennicutt/SINGS Team

Tables and illustrations are adapted, redrawn, or used by permission of the following authors and publishers:

Table 1.1: Robert Garrison and Toomas Karmo, *Observer's Handbook 2014*, with permission of the Royal Astronomical Society of Canada.

Table 2.1: *Astronomy: Fundamentals and Frontiers*, 3rd edition, by Robert Jastrow and Malcolm H. Thompson. Copyright © 1972, 1974, 1977 by Robert Jastrow (John Wiley & Sons, New York).

Table 3.1 (adapted); 11.2 (selected): *Allen's Astrophysical Quantities*, 4th edition. Copyright © 1999 by A. N. Cox, ed. (Springer-Verlag, New York).

Table 6.2 (adapted): *Realm of the Universe*, by George O. Abell. Copyright © 1964, 1969, 1973, 1980, by Holt, Rinehart and Winston, Inc. Copyright © 1976 by George O. Abell. Used by permission of Holt, Rinehart and Winston, Inc.

Tables 8.2 and 8.3: (selected) National Aeronautics and Space Administration public information. Updates by National Space Science Data Center, URL: http://nssdc.gsfc.nasa.gov

Tables 10.2 and 10.3: with permission from *Solar Eclipses: 1996–2020, 2021–2030* and *Lunar Eclipses: 1996–2020*, and *2021–2030* by Fred Espenak, NASA/Goddard Space Flight Center.

Tables 8.3 and 11.1: (selected) Brian G. Marsden, Smithsonian Astrophysical Observatory.

Appendix 5: Alan Batten, *Observer's Handbook 2009*, with permission of the Royal Astronomical Society of Canada.

TO THE READER

Astronomy is a user-friendly guide for beginners. Chapters make it easy for you to quickly learn the main topics of a college-level course. Sections clarify basic principles and contemporary advances. The Index enables you to look up concepts, definitions, facts, and famous astronomers, fast.

You can use the book alone as a handy reference or with a conventional textbook, Internet-based distance-learning course, homeschooling computer software, or telescope manual.

PARTICULARLY USEFUL FEATURES

- Website addresses throughout for the best astronomy online.
- Mathematics is not required.
- Line art makes technical ideas obvious.
- Star and Moon maps for fun stargazing.
- Up-to-date, accurate star, constellation, and astronomical data.
- Popular sky targets for hobby telescopes.
- Tips for hands-on, active learning.
- Objectives, reviews, and self-tests to monitor your progress.

WHAT'S NEW IN THE EIGHTH EDITION?

While keeping its successful self-teaching format, this eighth edition incorporates website addresses for spectacular color images. The entire book was revised to include *revolutionary* discoveries and the *best* suggestions from many readers and educators who profitably used prior editions.

Frontier *twenty-first-century* research into black holes, active galaxies and quasars, searches for life in space, origin and structure of our universe, and the newest ground and space telescopes are described.

Websites with daily astro-news and space scenes never before viewed by humans are specified. Labeled drawings of the Keck Telescope, Fermi Gamma Ray Observatory, and Hubble Space Telescope data path clarify space technology. New art illustrates fundamental concepts, such as the electromagnetic spectrum, phases of the Moon, planet orbits, and H-R diagrams.

STUDY AIDS

A list of objectives for each chapter tells you instantly what information is contained there. The first time a new term is introduced, it appears in bold type and is defined. Topics in each chapter are presented in short, numbered sections. Each section contains new information and usually asks you to answer a question or asks you to suggest an explanation, analyze, or summarize as you go along. You will always see the answer to the question right after you have answered it. If your answer agrees with the book's, you understand the material and are ready to proceed to the next section. If it does not, you should review some previous sections to make sure you understand the material before you proceed.

A self-test at the end of each chapter lets you find out fast how well you understand the material in the chapter. You may test yourself right after completing a chapter, or you might take a break and then take the self-test as a review before beginning a new chapter. Compare your answers with the book's. If your answers do not agree with the printed ones, review the appropriate sections (listed next to each answer).

USEFUL RESOURCES

Sources of excellent print and online astronomy materials, activities, and references are included in the Useful Resources section. Here you will also find a list of other books for stargazers of all ages by the author, Dinah L. Moché, Ph.D.

The author and publisher have tried to make this book accurate, up-to-date, enjoyable, and useful for you. It has been read by astronomers and many students, hobbyists, and educators who have contributed helpful suggestions during the preparation of the final manuscript. If while reading the book you have suggestions to improve it for future readers or for an author's visit, please let the author know! You can e-mail Dinah L. Moché, Ph.D., drdinah@spacelady.com, or visit her website. ▶www.spacelady.com◀

ACKNOWLEDGMENTS

I am especially grateful to my numerous students and lecture audiences and to readers of earlier editions of *Astronomy* whose questions and comments shaped the eighth edition.

Special thanks for always enthusiastically sharing the wonder and excitement of space with me to:

My home galaxy of stars, Mollie and Bertram A. Levine; Elizabeth, Stephen, Lucy, Benjamin, Robert, and Melanie Schwartz; and Rebecca, Richard, Cynthia, Jessica, Caroline, and Amanda Kahlenberg.

The National Science Foundation Faculty Fellowship in Science awarded to me made possible advanced studies in astronomy.

Thanks for the Eighth Edition go to: Rick Fienberg, Inge Heyer, Laurence A. Marschall, American Astronomical Society; Joy Pollard, Peter Michaud, Gemini Observatory; David Aguilar, Harvard Smithsonian Center for Astrophysics; Francis J. Reddy (GSFC), Wade E Sisler (JPL), Whitney B. Clavin (JPL) of NASA; Kathie Coil, National Optical Astronomy Observatory; President Diane Call, Raul Armendariz, Tak D. Cheung, M. Chantale Damas, Sunil Dehipawala, Francesca R. Gianferrara, Wenli Guo, Todd Holden, Alexei Kisselev, Dimitrios Kokkinos, David H. Lieberman, Paul Marchese, Charles Neuman, Arkadiy Portnoy, Charles Prancl, Vazgen Shekoyan, Winfield Sylvester, George Tremberger, Queensborough Community College of the City University of New York; Carlos Yascaribay, Queens College of CUNY; David M. F. Chapman, Royal Astronomical Society of Canada; Lynn Cominsky, Aurore Simonnet, Sonoma State University; Cheryl S. Gundy, Space Telescope Science Institute; Gordon Squires, Thirty Meter Telescope; Christina Roth, Jandra Sutton, Angie Lithgow, Laura Cusack, Turner Publishing Company; Thomas Jarrett, University of Capetown/Caltech; and Richard Nayman, Westchester PC Users Group.

I appreciate the continued encouragement and support of those who contributed to earlier editions.

Seventh Edition: Stephen P. Maran, American Astronomical Society;

T. H. Jarrett (IPAC) and J. Davy Kirkpatrick, California Institute of Technology; Peter B. Stetson, Dominion Astrophysical Observatory; John Pazmino, Federal Energy Regulatory Commission; Peter Michaud, Gemini Observatory; Laurence A. Marschall, Gettysburgh College; Kimberly Kowal Arcand, Harvard Smithsonian Center for Astrophysics; Megan Watzke, Harvard Smithsonian Center for Astrophysics; Antoinette Beiser, Lowell Observatory; Deidre Hunter, Lowell Observatory; Dawn Myers (GSFC), William Steigerwald (GSFC), Cheryl Gundy (STScI), Eric L. Winter (IPAC), NASA; Dave Finley, National Radio Astronomy Observatory; Ruth A. Kneale, National Solar Observatory; Santi Cassisi, Osservatorio di Teramo; President Eduardo Marti, Tak D. Cheung, Thomas P. Como, Alex L. Flamholz, Francesca R. Gianferrara, Todd M. Holden, Alec Kisselev, Liza Larios, David H. Lieberman, Paul J.Marchese, Bruce Naples, Charles P. Neuman, Charles Prancl, Ralph Romanelli, Robert Taylor, George Tremberger Jr., Queensborough Community College of the City University of New York; Seth Shostak, SETI Institute; Aurore Simonnet, Sonoma State University; Lynn Cominsky, Sonoma State University; Patrick Kelly, The Royal Astronomical Society of Canada; Barbara J. Anthony-Twarog, University of Kansas; Don Vandenberg, University of Victoria; Stephen Kippur, Dean Karrel, Kitt Allan, Eric Nelson, Constance Santisteban, Ellen Wright, Camille Acker, Megan Burke, and Barbara Mele, John Wiley & Sons, Inc.; and Robert Benjamin, University of Wisconsin.

Sixth Edition: Davy Kirkpatrick, California Institute of Technology; Marion Schmitz, California Institute of Technology; Peter Michaud, Gemini Observatory; Laurence A. Marschall, Gettysburg College; Robert Kirshner, Harvard University; David Aguilar, Harvard-Smithsonian Center for Astrophysics; Brian G. Marsden, Harvard-Smithsonian Center for Astrophysics; Gareth V. Williams, Harvard-Smithsonian Center for Astrophysics; Holland Ford, Johns Hopkins University; Jim Lochner, NASA Goddard Space Flight Center; Stephen Maran, NASA Goddard Space Flight Center; Rajiv Gupta, Royal Astronomical Society of Canada; Lynn Cominski, Sonoma State University; Cheryl Gundy, Space Telescope Science Institute; Zoltan Levay, Space Telescope Science Institute; Ray Villard, Space Telescope Science Institute; Alex Filippenko, University of California/Berkeley; Harold Epps, University of California/Santa Cruz.

Fifth Edition: Joseph F. Veverka, Cornell University; Robert Garrison, David Dunlap Observatory; Alan Batten, Dominion Astrophysical Observatory; Peter Michaud, Gemini Observatory; Francois Spite, IAU; Gerard Helferich, John Wiley & Sons, Inc.; Christopher Jackson and Diana C. Madrigal, John Wiley & Sons; Michael Arida, Fred Espenak, Stephen P. Maran, Wayne Warren (GSFC), Alan Chamberlain, Mary Beth Murrill, Jane Platt (JPL), and Cheryl Gundy (STScI), NASA; David G. Finley, National Radio Astronomy Observatory; Roy L. Bishop, Royal Astronomical Society of Canada; Brian Marsden, Smithsonian Astrophysical Observatory; Geoff Chester, U.S. Naval

Observatory; Harry Shipman, University of Delaware; and Helene Dickel, University of Illinois.

Fourth Edition: Steve Maran, American Astronomical Society; Maria Pallante, Authors Guild; Bob Finn, California Institute of Technology; Richard Dannay, Esq.; Pat Peterson, de Grummond Collection, University of Southern Mississippi; Carol R. Leven, Freelance Administrator; Laurence A. Marschall, Gettysburg College; Nicholas L. Johnson, Kaman Sciences Corporation; Mary Beth Murrill, W. M. Keck Observatory; Keith Mordoff, Lockheed Missiles & Space Company, Inc.; Richard Jackson, Bill Santoro, Joe Schank, Mamaroneck Post Office; Constance Moore, Althea Washington (Headquarters), Alan S. Wood, Kimberly Lievense, Sharon Miller, Mary Hardin, Ed McNevin, Jurrie van der Woude, Gil Yanow (JPL), Charles Borland, Billie A. Deason, Lisa Vazquez (JSC), Allen Kenitzer (MSFC), Ray Villard (STSI), National Aeronautics and Space Administration; Emma Hardesty, Karie Myers, National Optical Astronomy Observatories; Director Paul A. Vanden Bout, Patrick C. Crane (VLA), Pat Smiley, National Radio Astronomy Observatory; Array; Roy Bishop, *Observer's Handbook*; Gloria Lubkin, *Physics Today*; Jacqueline Mitton, Royal Astronomical Society (U.K.); David Okerson, Science Applications International Corporation; George Lovi, *Sky and Telescope* columnist; Preston J. Campbell, TRW Federal Systems Division; John Percy, University of Toronto; Jay Pasachoff, Williams College.

Third Edition: I. Robert, Victor and Esther Rozen; Jack Flynn, Andrew Fraknoi, Juliana Ver Steeg, Astronomical Society of the Pacific; Director Sidney Wolff, Carl A. Posey, and Jeff Stoner, Kitt Peak National Observatory; Elyse Murray, Bernard Oliver, and Charles Seeger (Ames), Donald K. Yeomans (JPL), NASA; Ronald Ekers, Arnold H. Rots, and Don L. Swann, NRAO/VLA; Tobias Owen, SUNY/Stony Brook; Larry Esposito, University of Colorado; and Paul W. Hodge, University of Washington.

Second Edition: Lloyd Motz and Chien Shiung Wu, Columbia University; Harry L. Shipman, University of Delaware; Frank E. Bristow (JPL), Les Gaver, David W. Garrett, Curtis M. Graves, William D. Nixon (Headquarters), Peter W. Waller (Ames), and Terry White (JPL), NASA; Janet K. Wolfe, National Air and Space Museum; Richard W. West, NSF; Henry D. Berney, Thomas Como, Donald Cotten, Julius Feit, Sheldon E. Kaufman, Valdar Oinas, Robert Taylor, and Kurt R. Schmeller, Queensborough Community College of CUNY; and Arnold A. Sterassenburg, SUNY/Stony Brook.

CONTENTS

LIST OF TABLES

INTRODUCTION: COSMIC VIEW

Strange is our situation here upon Earth.
Each of us comes for a short visit, not knowing why,
Yet sometimes seeming to divine a purpose.

Albert Einstein (1879–1955)

On a clear night in a place where the sky is really dark, you can see about 2000 stars with your unaided eye. You can look trillions of kilometers into space and peer thousands of years back into the distant past.

As you gaze at the stars you may wonder: What is the pattern or meaning of the starry heavens? What is my place in the vast cosmos? You are not alone in asking these questions. The beauty and mystery of space have always fascinated people.

Astronomy is the oldest science—and the newest. Exciting discoveries are being made today with the most sophisticated tools and techniques ever available. Yet you can still make important contributions.

This book will teach you the basic concepts of astronomy and space exploration. You will more fully enjoy observing the stars as your knowledge and understanding grow. You will be better able to surf the web and to read more on topics that intrigue you, from ancient astronomy to the latest astrophysical theories and spaceflights.

As you teach yourself astronomy, refer to these tools:

The Star maps and Moon map at the back of this book. These special, easy-to-read maps will help you locate and identify particularly interesting objects in the sky.

Simple activities you can do that demonstrate a basic idea.

Participate in astronomical research from your computer or mobile device.

▶ Internet link to spectacular images and new reports.

Now, begin reading about the enormous tracts of space and time we call the universe, and stretch your mind!

Our home is planet Earth, a rocky ball about 13,000 km (8000 miles) in diameter suspended in the vastness of space-time (Figure I.1).

Figure I.1. Earth photographed from space. Sunshine dramatically spotlights Earth's blue ocean, reddish-brown land masses, and white clouds from the Mediterranean Sea area to the Antarctica polar ice cap.

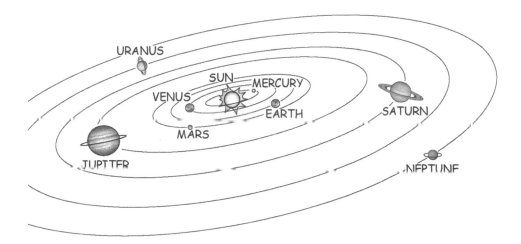

Figure I.2. Planets orbiting the Sun in our solar system. (Drawing not to scale.)

Earth belongs to our solar system (Figure I.2). Our solar system consists of one star—our Sun—plus planets, moons, small solar system bodies, and dust particles, all of which revolve around the Sun. Our solar system is more than 15 trillion km (9 trillion miles) across.

The Sun and the solar system are located in one of the great spiral arms of the Milky Way Galaxy (Figure I.3). Our immense Milky Way Galaxy

Sun with its planets
and other bodies

Figure I.3. Our solar system in the Milky Way Galaxy.

includes over 200 billion stars plus interstellar gas and matter, both luminous and dark, all revolving around the center. The Milky Way Galaxy is about 100,000 light-years across. (One light-year is practically 10 trillion km, or 6 trillion miles.)

Our Milky Way Galaxy is only one of billions of galaxies that exist to the edge of the observable universe, some 14 billion light-years away (Figure I.4).

Figure I.4. Nearly 5,500 distant galaxies in a patch of sky a small fraction as big as the full Moon, in the constellation Fornax. Each galaxy includes billions of stars.

1

UNDERSTANDING THE STARRY SKY

And that inverted bowl we call the Sky
Where under crawling coop't we live and die
Lift not your hands to it for help—for It
As impotently rolls as you and I.

Rubáiyát of Omar Khayyám (1048–1131)

Objectives

☆ Locate sky objects by their right ascension and declination on the celestial sphere.

☆ Identify some bright stars and constellations visible each season.

☆ Explain why the stars appear to move along arcs in the sky during the night.

☆ Explain why some different constellations appear in the sky each season.

☆ Explain the apparent daily and annual motions of the Sun.

☆ Define the zodiac.

☆ Describe how the starry sky looks when viewed from different latitudes on Earth.

☆ Define a sidereal day and a solar day, and explain why they differ.

☆ Explain how astronomers classify objects according to their apparent brightness (magnitude).

☆ Explain why the polestar and the location of the vernal equinox change over a period of thousands of years.

1.1 STARGAZER'S VIEW

On a clear, dark night the sky looks like a gigantic dome studded with stars. We can easily see why the ancients believed that the starry sky was a huge sphere turning around Earth.

Today we know that stars are remote, blazing suns racing through space at different distances from Earth. The Earth **rotates**, or turns, daily around its **axis** (the imaginary line running through its center between the North and South Poles).

But the picture of the sky as a huge, hollow globe of stars that turns around Earth is still useful. Astronomers call this fictitious picture of the sky the **celestial sphere**. "Celestial" comes from the Latin word for heaven.

Astronomers use the celestial sphere to locate stars and galaxies and to plot the courses of the Sun, Moon, and planets throughout the year. When you look at the stars, imagine yourself inside the celestial sphere looking out (Figure 1.1).

Why do the stars on the celestial sphere appear to move during the night when you observe them from Earth? _____

_____

Answer: Because the Earth is rotating on its axis inside the celestial sphere.

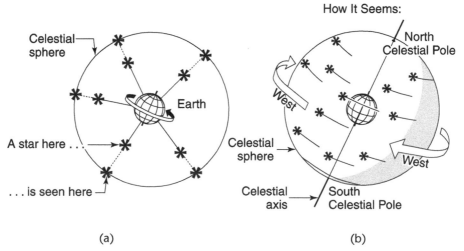

(a)   (b)

Figure 1.1. (a) To a stargazer on Earth, all stars appear equally remote. (b) We picture the stars as fixed on a celestial sphere that spins westward daily (opposite to Earth's actual rotation).

1.2 CONSTELLATIONS

It is fun to go outside and see a young blue-white star or a dying red giant star in the sky right after you read about them. You may think you will never be able to tell one star from another when you begin stargazing, but you will.

The removable star maps at the back of this book have been drawn especially for beginning stargazers observing from around 40°N latitude. (They should be useful to new stargazers throughout the midlatitudes of the northern hemisphere.)

Stars appear to belong to groups that form recognizable patterns in the sky. These star patterns are called **constellations**. Learning to identify the most prominent constellations will help you pick out individual stars.

The 88 constellations officially recognized by the International Astronomical Union are listed in Appendix 1. Famous ones that shine in these latitudes are shown on your star maps. Their Latin names and the names of **asterisms**, or popular unofficial star patterns, are printed with the initial letters capitalized.

Thousands of years ago people named the constellations after animals, such as Leo the Lion (Figure 1.2), or mythological characters, such as Orion the Hunter (Figure 5.1). More than 2000 years ago the ancient Greeks recognized 48 constellations.

Modern astronomers use the historical names of the constellations to refer to 88 sections of the sky rather than to the mythical figures of long ago. They refer to constellations in order to locate sky objects. For instance, saying that Mars is in Leo helps locate that planet, just as saying that Houston is in Texas helps locate that city.

Look over your star maps. Notice that the dashed line indicates the **ecliptic**, the apparent path of the Sun against the background stars. The 12

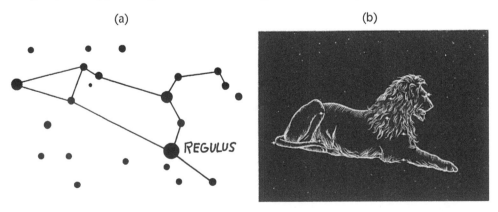

(a)   (b)

REGULUS

Figure 1.2. Constellation Leo is best seen in early spring when it is high in the sky. (a) Brightest star Regulus marks the lion's heart, a sickle of stars his mane, and a triangle of stars his hindquarters and tail. (b) Leo the Lion.

constellations located around the ecliptic are the constellations of the zodiac, whose names are familiar to horoscope readers.

List the 12 constellations of the zodiac. _____

_____

Answer: Pisces, Aries, Taurus, Gemini, Cancer, Leo, Virgo, Libra, Scorpius, Sagittarius, Capricornus, Aquarius.

1.3 CIRCUMPOLAR CONSTELLATIONS ★

Study your star maps carefully. You will notice that several **circumpolar con tlla tion** , near the north celestial pole (marked POLE +), appear on all four maps.

These are **north circumpolar constellations**, visible above the northern horizon all year long at around 40°N latitude (Figure 1.3). At this latitude, the south celestial pole and nearby **south circumpolar constellations** do not rise above the horizon any night of the year.

List the three circumpolar constellations closest to Polaris (the North Star) and sketch their outlines. _____

_____

_____

_____

Answer: Three circumpolar constellations that you should be able to pick out on the star maps are Cassiopeia, Cepheus, and Ursa Minor. After you know their outlines, try to find them in the sky above the northern horizon. *Note*: At latitude 40°N or higher, Ursa Major and Draco are also circumpolar.

1.4 HOW TO USE THE STAR MAPS ★

You can use the star maps outdoors to *identify* the constellations and stars you see in the night sky and to *locate* those you want to observe.

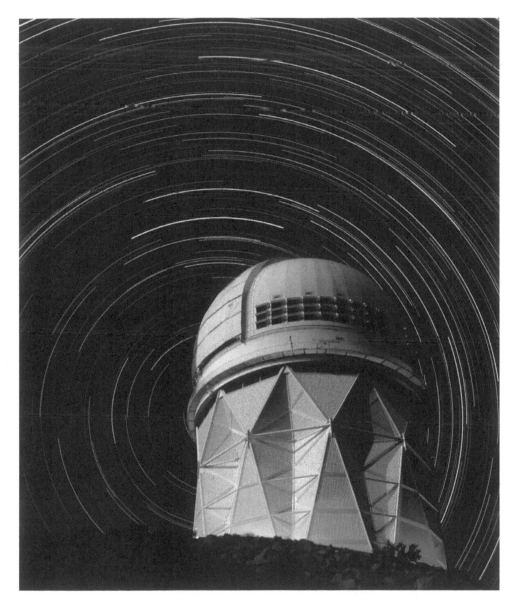

Figure 1.3. A time exposure taken with a camera aimed at the north celestial pole over the U.S. Kitt Peak National Observatory shows star trails that mirror Earth's actual rotation. Kitt Peak is a 2100-m-high (6900-ft-high) site about 30 km (50 miles) outside of Tucson, Arizona. ▶www.noao.edu/kpno◀

Choose the map that pictures the sky at the month and time you are stargazing. Turn the map so that the name of the compass direction you are facing appears across the bottom. Then, from bottom to center, your star map pictures the sky as you are viewing it from your horizon to the point directly over your head.

For example, if you are facing north about 10:00 P.M. in early April, turn the spring skies map so that the word NORTH is at the bottom. From the horizon up, you may observe Cassiopeia, Cepheus, the Little Dipper in Ursa Minor, and the Big Dipper in Ursa Major.

Name a prominent constellation that shines in the south at about 8:00 P.M. in early February. _____

Answer: Orion.

1.5 HOW TO IDENTIFY CONSTELLATIONS

The constellations above the southern horizon parade by during the night and change with the seasons. Turn each map so that the word SOUTH is at the bottom. Use your star maps to identify the most prominent constellations that shine each season (such as Leo in the spring and Orion in the winter).

Identify and sketch three constellations that you can see this season.

_____

_____

Answer: Your answer will depend on the season. For example, if you are reading this book in the spring, you might choose Leo, Virgo, and Boötes.

1.6 STAR NAMES

Long ago, more than 50 of the brightest stars were given proper names in Arabic, Greek, and Latin. The names of bright or famous stars to look for are printed on your star maps with the initial letters capitalized.

Today astronomers use alphabets and numerals to identify hundreds of thousands of stars. They refer to each of the brightest stars in a constellation by a Greek letter plus the Latin genitive (possessive) form of the constellation name. Usually the brightest star in a constellation is α, the next brightest is β, and so on. (The Greek alphabet is listed in Appendix 3.) Thus, Regulus is called α Leonis, or the brightest star of Leo. Fainter stars, not shown on your maps, are identified by numbers in star catalogs.

In a built-up metropolitan area you can see only the brightest stars. When you are far from city lights and buildings and the sky is very dark and clear, you can see about 2000 stars with your unaided eye.

Name the three bright stars that mark the points of the famous Summer Triangle. Refer to your summer skies map. _____   _____

Answer: Vega, Deneb, and Altair. Look for the Summer Triangle overhead during the summer.

1.7 BRIGHTNESS

Some stars in the sky look brighter than others. The **apparent magnitude** of a sky object is a measure of its observed brightness as seen from Earth. Stars may look bright because they send out a lot of light or because they are relatively close to Earth.

In the second century B.C., the Greek astronomer Hipparchus divided the visible stars into six classes, or magnitudes, by their relative brightness. He numbered the magnitudes from 1 (the brightest) through 6 (the least bright).

Modern astronomers use a more precise version of the ancient classifying system. Instead of judging brightness by the eye, they use an instrument called a **photometer** to measure brightness. Magnitudes for the brightest stars are negative—the brightest night star, Sirius, measures –1.44. Magnitudes range from –26.75 for the Sun to about +31 for the faintest objects observed in a space telescope. A difference of 1 magnitude means a brightness ratio of about 2.5.

Magnitudes are shown on your star maps and in Table 1.1. For example, we receive about 2.5 times as much light from Vega, a star of magnitude 0, as we do from Deneb, a star of magnitude 1, and about 6.3 times as much light as from Polaris, of magnitude 2. (Magnitudes are discussed further in Section 3.14.)

What do astronomers mean by "apparent magnitude"? _____

Answer: How bright a sky object looks.

1.8 LOCATION ON EARTH

The more you understand about stars and their motions, the more you will enjoy stargazing. A **celestial globe** helps you locate sky objects as a **terrestrial** (Earth) **globe** helps you locate places on Earth.

Remember how Earth maps work. We picture the Earth as a sphere and draw imaginary guidelines on it. All distances and locations are measured from two main reference lines, each marked 0°. One line, the **equator**, is the great circle halfway between the North and South Poles that divides the globe into halves. The other line, the **prime meridian**, runs from pole to pole through Greenwich, England.

Imaginary lines parallel to the equator are called **latitude** lines. Those from pole to pole are called **longitude** lines, or **meridians**. You can locate any city on Earth if you know its coordinates of latitude and longitude. Distance on the terrestrial sphere can be measured by dividing the sphere into 360 sections, called **degrees** (°). (Angular measure is defined in Appendix 3.)

Refer to the globe in Figure 1.4. Identify the equator; prime meridian; 30°N latitude line; and 30°E longitude line. (a) _____ ; (b) _____ ; (c) _____ ; (d) _____

Answer: (a) 30°N; (b) 30°E; (c) equator; (d) prime meridian.

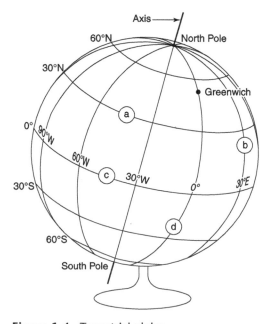

Figure 1.4. Terrestrial globe.

1.9 CELESTIAL COORDINATES

Astronomers draw imaginary horizontal and vertical lines on the celestial sphere similar to the latitude and longitude lines on Earth. They use celestial coordinates to specify directions to sky objects.

The **celestial equator** is the projection of the Earth's equator out to the sky. Angular distance above or below the celestial equator is called **declination (dec)**. Distance measured eastward along the celestial equator from the zero point, the **vernal equinox**, is called **right ascension (RA)**. Right ascension is commonly measured in hours (h), with $1^h = 15°$.

Just as any city on Earth can be located by its coordinates of longitude and latitude, any sky object can be located on the celestial sphere by its coordinates of right ascension and declination.

Give the location of the star shown in Figure 1.5. _____

Answer: 20^h RA, 30°N declination.

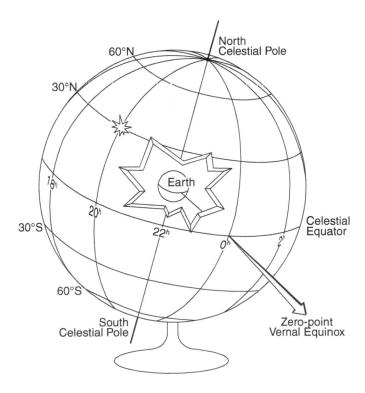

Figure 1.5. Celestial globe.

1.10 LOCATION ON THE CELESTIAL SPHERE

Every star has a location on the celestial sphere, where it appears to be when sighted from Earth. The right ascension and declination of stars for a standard **epoch**, or point of time selected as a fixed reference, change little over a period of many years. They can be read from a celestial globe, star atlas, or computer software. (See Table 1.1, for example. You'll be referring to this table when the information it contains is discussed in later chapters.)

TABLE 1.1 The Brightest Stars

	Star Name	Right Ascension h	m	Declination °	′	Apparent Magnitude	Spectral Class	Distance (ly)	Absolute Magnitude
Sun		—	—	—	—	–26.75	G	8 lm	4.8
Sirius	α Canis Majoris	06	45.8	–16	44	–1.44	A	8.6	1.5
Canopus	α Carinae	06	24	–52	42	–0.62	A	310	–5.4
Arcturus	α Bootis	14	16.3	+19	06	–0.05	K	38	–0.6
Rigil Kentaurus	α Centauri	14	40.6	–60	54	–0.01	G	4.3	4.2
Vega	α Lyrae	18	37.4	+38	48	0.03	A	25	0.6
Capella	α Aurigae	05	17.8	+46	01	0.08	G	45	–0.8
Rigel	ß Orionis	05	15.2	–08	11	0.18	B	900	–6.6
Procyon	α Canis Minoris	07	40.1	+05	11	0.40	F	11.5	2.8
Achernar	α Eridani	01	38.3	–57	10	0.45	B	140	–2.9
Betelgeuse	α Orionis	05	56	+07	25	0.45	M	500	–5.0
Hadar	ß Centauri	14	04.9	–60	27	0.58	B	400	–5.5
Altair	α Aquilae	19	51.5	+08	54	0.76	A	16.7	2.1
Aldebaran	α Tauri	04	36.8	+16	32	0.87	K	67	–0.8
Spica	α Virginis	13	26	–11	14	0.98	B	250	–3.6
Antares	α Scorpii	16	30	–26	28	1.06	M	600	–5.8
Pollux	ß Geminorum	07	46.2	+27	59	1.16	K	33.84	1.1
Formalhaut	α Piscis Austrini	22	58.4	-29	33	1.17	A	25.1	1.6
Deneb	α Cygni	20	41.9	+45	20	1.25	A	1400	–7.5
Acrux	α Crucis	12	27.4	–63	11	1.25	B	320	–4.0
Becrux	ß Crucis	12	48.6	–59	46	1.25	B	300	–4.0

Note: Magnitudes are visual magnitudes, measured over visible wavelengths.

Abbreviations
Right Ascension: h = hours; m = minutes of time
Declination: ° = degrees; ′ = minutes of arc
Distance: ly = light-year; lm = light-minute

The locations of the Sun, Moon, and planets on the celestial sphere change regularly. You can find their monthly positions, rise and set times, and other practical data in current astronomical publications, websites, software, and apps (see Useful Resources). The U.S. Naval Observatory provides data and the official U.S. time. ▶http://aa.usno.navy.mil◀

Explain why in any given era the stars may be found at practically the same coordinates on the celestial sphere, while the Sun, Moon, and planets change their locations regularly. _____

_____

_____

_____

_____

Answer: The stars are too far from Earth for the unaided eye to see them move even though they are traveling many kilometers per second in various directions. The Sun, Moon, and planets are much closer to Earth. We see them move relative to the distant stars.

1.11 LOCAL REFERENCE LINES

Lines of declination and right ascension are fixed in relation to the celestial sphere and move with it as it rotates around an observer. Other useful reference lines relate to the local position of each observer and stay fixed with the observer while sky objects pass by.

At your site, the **zenith** is the point on the celestial sphere directly over your head. The **celestial horizon** is the great circle on the celestial sphere 90° from your zenith. Although the celestial sphere is filled with stars, you can see only those that are above your horizon. The **celestial meridian** is the great circle passing through your zenith and the north and south points on your horizon. Only half of the celestial meridian is above the horizon.

Refer to Figure 1.6. Identify the stargazer's zenith; celestial horizon; and celestial meridian. (a) _____ ; (b) _____ ; (c) _____

Answer: (a) Zenith; (b) meridian; (c) horizon.

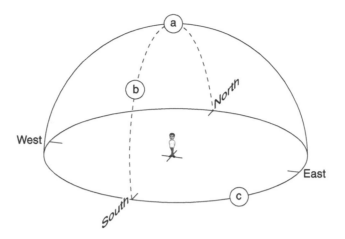

Figure 1.6. A stargazer's local reference lines.

1.12 CELESTIAL MERIDIAN 💡

Go outside and trace out your zenith, celestial horizon, and celestial meridian by imagining yourself, like that stargazer, at the center of the huge celestial sphere.

If possible, try this on a clear, dark, starry night. Face south. Observe the stars near your celestial meridian several times during the night. Describe what you observe. _____

_____

_____

Answer: The stars move from east to west and **transit**, or cross, your celestial meridian. This is because of the Earth's rotation from west to east. A star **culminates**, or reaches its highest altitude, when it is on the celestial meridian.

1.13 LATITUDE AND STARGAZING 💡

The stars that appear above your horizon and their paths across the sky depend on your latitude on Earth. The sky looks different from different latitudes (Figure 1.7).

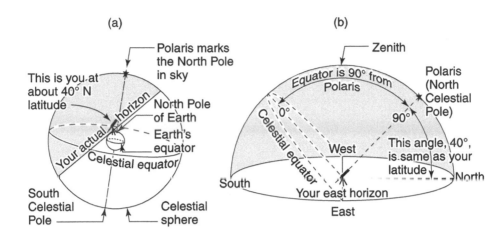

(a)

(b)

Figure 1.7. Local orientation of the celestial sphere at 40°N latitude. (a) View from a fictitious spot on the outside. (b) Stargazer's view.

If you could look at the sky from the North Pole and then from the South Pole you would see completely different stars. The Earth cuts your view of the celestial sphere in half.

You can determine how the celestial sphere is oriented with respect to your horizon and zenith at any place on Earth. In the northern hemisphere, the north celestial pole is located above your northern horizon at an altitude equal to your latitude. Polaris, the polestar, or North Star, is less than one degree away from the north celestial pole and marks the position of the pole in the sky. The declination circle that is numerically equal to your latitude passes through your zenith. In the southern hemisphere, the south celestial pole is located above your southern horizon at an altitude equal to your latitude. It is not marked by a polestar.

Where would you look for the North Star if you were at each of the following locations: (a) the North Pole? _____ (b) the equator? _____ (c) 40°N latitude? _____ (d) your home? _____

Answer: (a) At your zenith; (b) on your horizon; (c) 40° above your northern horizon; (d) at an altitude above your northern horizon equal to your home latitude.

1.14 APPARENT DAILY MOTION OF THE STARS

The stars appear to move in **diurnal circles**, or daily paths, around the celestial poles when you observe them from the spinning Earth.

Although the North Star, Polaris, is not a very bright star, it has long been important for navigation. Closest to the north celestial pole, it is the only star that seems to stay in the same spot in the sky. You can find Polaris by following the "pointer stars," Dubhe and Merak, in the bowl of the Big Dipper in the constellation Ursa Major (Figure 1.8).

Since the celestial poles are at distinct altitudes in the sky at distinct latitudes, the part of a star's diurnal circle that is above the horizon is different at different latitudes on Earth (Figure 1.9).

For example, if you stargaze at 40°N latitude, about the latitude of Denver, Colorado, U.S., you will see these stars (Figure 1.9): (1) Stars within 40° (your latitude) of the north celestial pole (those stars between +50° and +90° declination) are always above your horizon. These stars that never set—such as the stars in the Big Dipper—are **north circumpolar stars**. (2) Stars that are within 40° (your latitude) of the south celestial pole never appear

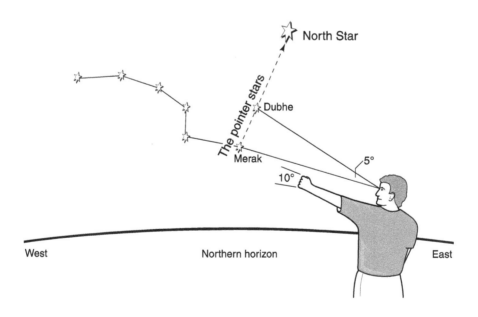

Figure 1.8. The "pointer" stars, Dubhe and Merak, in the bowl of the Big Dipper lead you to the North Star, Polaris. The angular distance between these pointer stars is about 5° on the celestial sphere. A fist at arm's length marks about 10°. These examples will help you judge other angular distances in the sky.

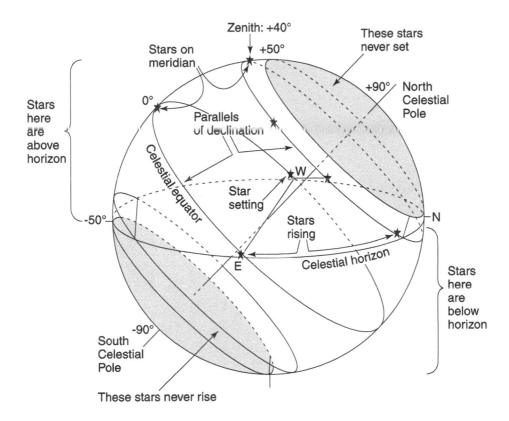

Figure 1.9. The sky from 40°N latitude. The north celestial pole is 40° above the northern horizon, and the celestial sphere rotates around it. Parallels of declination mark the stars' diurnal circles.

above your horizon. These stars that never rise—such as the stars in the constellation Crux, the Southern Cross—are **south circumpolar** stars. (3) The other stars, in a band around the celestial equator, rise and set. Those stars that are located at 40°N declination (equal to your latitude) pass directly across your zenith when they cross your celestial meridian.

Assume you are stargazing at 50°N latitude, about the latitude of Vancouver, Canada. Refer to Table 1.1 for the declinations of the bright stars Capella, Vega, and Canopus. Which of these stars will be above your horizon:

(a) always? _____ (b) sometimes? _____ (c) never? _____

Answer: (a) Capella (+46°00′ declination). Stars within 50° of the north celestial pole (between +40° and +90° declination) are always above the horizon. (b) Vega (+38°47′ declination). This star rises and sets. (c) Canopus (−52°42′ declination) is within 50° of the south celestial pole (between −40° and −90° declination).

1.15 UNUSUAL VIEWS

Describe how the diurnal circles of the stars would look if you were stargazing at (a) the North Pole and (b) the equator. Explain your answer. *Tip:* Remember that the celestial sphere rotates around the celestial poles. (a) _____

_____

_____

(b) _____

_____

_____

Answer: (a) All stars would seem to move along circles around the sky parallel to your horizon. The celestial sphere rotates around the north celestial pole, which is located at your zenith at the North Pole. (b) All stars would seem to rise at right angles to the horizon in the east and set at right angles to the horizon in the west. The celestial sphere rotates around the celestial poles, which are located on your horizon at the equator.

1.16 APPARENT ANNUAL MOTION OF THE STARS

The appearance of the sky changes during the night because of Earth's rotation. It also changes slowly from one night to the next.

Every night the stars appear a little farther west than they did at the same time the night before. A star rises about 4 minutes earlier each evening. Four minutes a day for 30 days adds up to about 2 hours a month. If a star is above the horizon during the daytime, the bright Sun will obscure it from view.

Thus the stars that shine in your sky at a particular time change noticeably from month to month and from season to season. In 12 months, that 4 minutes a day adds up to 24 hours. After a year, the starry sky looks the same again.

The change in the appearance of the sky with the change in seasons is due to the motion of the Earth around the Sun. The Earth **revolves**, or travels around, the Sun every year.

Picture yourself riding on Earth around the Sun, inside the celestial sphere, looking straight out. As Earth moves along in its orbit, your line of sight points toward different stars in the night sky. During a whole year you view a full circle of stars.

(a) If a star is on your zenith at 9:00 P.M. on September 1, about what time will it be on your zenith on March 1? _____ (b) Will you be able to see it? _____ Explain your answer. _____

_____

Answer: (a) About 9:00 A.M. Stars rise about 2 hours earlier every month. (b) No. At that hour of the day the bright Sun obscures the distant stars from view.

1.17 THE ECLIPTIC

If the stars were visible during the day, you would see the Sun apparently move eastward among them during the year. The **ecliptic**, the apparent path of the Sun against the background stars, is drawn on sky globes and star maps for reference.

The band about 16° wide around the sky that is centered on the ecliptic is called the **zodiac**. Ancient astrologers divided the zodiac into 12 constellations, or **signs**, each taken to extend 30° of longitude (see Appendix 3). The zodiac has attracted special attention because the Moon and planets, when they appear in the sky, also follow paths near the ecliptic through these 12 constellations (Figure 1.10).

What is the zodiac? _____

_____

Answer: A belt about 16° wide around the sky, centered on the ecliptic, containing 12 constellations.

1.18 APPARENT ANNUAL MOTION OF THE SUN

The apparent easterly motion of the Sun among the stars is caused by the real revolution of Earth around the Sun. The Sun seems to move in a full circle around the celestial sphere every year.

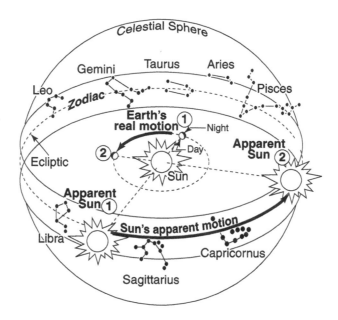

Figure 1.10. The Sun's apparent annual motion around the celestial sphere results from Earth's real motion around the Sun. As Earth orbits the Sun, different constellations of the zodiac appear in the night sky.

About how far does the Sun move on the ecliptic every day? *Tip:* Use the fact that the Sun moves 360° around the ecliptic in a year (about 365 days).

Answer: About 1°.

Solution:

$$\frac{360°}{365 \text{ days}} \cong 1° \text{ per day}$$

1.19 EARTH'S SEASONS

The Sun's path across the sky is highest in summer and lowest in winter. The altitude of the Sun above the horizon at noon varies during the year because Earth's axis is tilted to the plane of its orbit around the Sun (Figure 1.11).

Earth's equator remains tilted at about 23.5° to its orbital plane all year long. So as Earth travels around the Sun, the slant of the Earth–Sun line

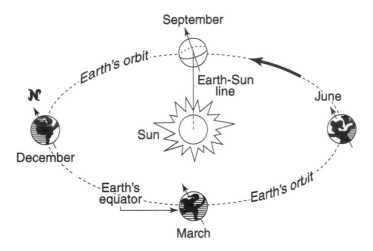

Figure 1.11. Because Earth's axis is tilted, each hemisphere gets varying amounts of Earth–Sun during the year as our planet orbits the Sun.

changes. Sunlight pours down to Earth from different angles during the year, causing the change of seasons as well as seasonal variations in the length of days and nights.

Refer to Figure 1.11. Is the northern hemisphere tipped toward or away from the Sun (a) in December? _____ (b) in June? _____

Answer: (a) Away from; (b) toward.

1.20 EQUINOXES AND SOLSTICES

You can determine what the Sun's apparent position in the sky will be on any given day by checking the ecliptic on a celestial globe or a flat sky map like the one in Figure 1.12.

The **vernal equinox**, which occurs about March 20, is the Sun's position as it crosses the celestial equator going north. It is the point on the celestial sphere chosen to be the 0^h of right ascension (see Section 1.9). The **autumnal equinox**, which occurs about September 23, is the Sun's position as it crosses the celestial equator going south. At the equinoxes, day and night are equal in length.

The **summer solstice**, which occurs about June 21, and the **winter solstice**, which occurs about December 21, are the most northern and most southern positions of the Sun during the year. At these times we have the longest and shortest days, respectively, in the northern hemisphere.

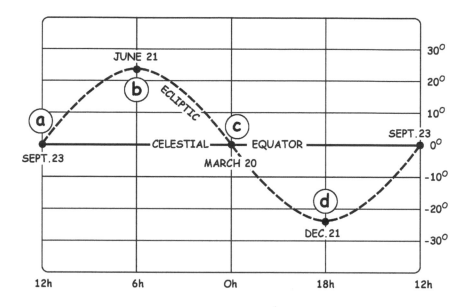

Figure 1.12. Flat sky map.

Refer to Figure 1.12. Identify the vernal equinox _____ ; autumnal equinox _____ ; summer solstice _____ ; and winter solstice _____ .

Answer: Vernal equinox (c); autumnal equinox (a); summer solstice (b); winter solstice (d).

1.21 SUN'S ALTITUDE

The Sun is never directly overhead for stargazers in the midlatitudes. On a given day, the maximum altitude of the Sun in your sky depends on its declination and your latitude.

Where would you have to stand on Earth to have the Sun pass directly across your zenith at the time of the (a) vernal equinox? _____ (b) summer solstice? _____ (c) autumnal equinox? _____ (d) winter solstice? _____

Answer: (a) Equator; (b) 23.5°N latitude (Tropic of Cancer); (c) equator; (d) 23.5°S latitude (Tropic of Capricorn).

1.22 OBSERVABLE EFFECTS OF EARTH'S MOTIONS

How do the motions of Earth in space cause noticeable changes in the appearance of the sky for an observer on Earth?_____

_____

_____

_____

_____

Answer: Your summary should include the following concepts: The starry sky changes during the night because of Earth's daily rotation. The visible stars change with the seasons because of Earth's annual revolution around the Sun. The Sun's apparent daily motion across the sky is due to Earth's real rotation. The Sun's apparent annual motion is due to Earth's real revolution.

1.23 THE DAY

Earth's rotation provides a basis for keeping time using astronomical observations. The **solar day** of everyday affairs measures the time interval of Earth's rotation using the Sun for reference. The **sidereal day** measures the time interval of Earth's rotation using the stars for reference.

A sidereal day is 23 hours, 56 minutes, 4 seconds long. It is the time interval required for a star to cross your meridian two times successively, or the time for Earth to complete one whole turn in space. A solar day is 24 hours long, the length of time required for two successive meridian transits by the Sun.

A solar day is about 4 minutes longer than a sidereal day because while Earth rotates on its axis it also moves along in its orbit around the Sun. Earth must complete slightly more than one whole turn in space before the Sun reappears on your meridian (Figure 1.13).

A clock that keeps sidereal time is useful for stargazing. In sidereal time, all stars return to their identical positions in the sky every 24 hours. So a star rises, transits the meridian, and sets at the same sidereal time all year long.

You can use celestial coordinates (see Table 1.1) to determine the sidereal time at any instant when you are stargazing. Local sidereal time is equal to the right ascension of stars on your meridian. For example, if you see brilliant Sirius transit, the sidereal time is 6 hours, 45 minutes.

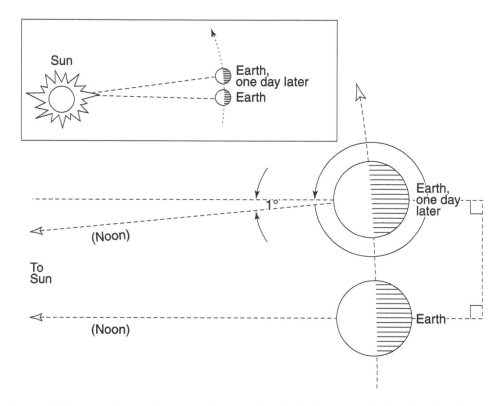

Figure 1.13. A solar day is longer than a sidereal day because during the time Earth rotates it also moves along its orbit around the Sun. In the interval from one noon to the next, Earth completes slightly more than one whole turn in space.

What motion of Earth causes the 4-minute difference between a sidereal and a solar day? _____

Answer: Earth's revolution around the Sun.

1.24 PRECESSION ★

Your star maps will be useful to you for the rest of your life. You may be interested to know, however, that they will finally go out of date hundreds of years from now.

Earth's axis of rotation shifts extremely slowly around a cone in space once about every 25,800 years. This slow motion of Earth's axis, caused mainly by the tug of the gravity of the Sun and Moon on Earth's equatorial bulge, is called **precession**.

Earth's axis always tilts 23.5° to its orbital plane, so precession causes the north celestial pole to circle among the stars. After thousands of years, the polestar changes (Figure 1.14).

The vernal equinox, the zero point of right ascension, drifts westward around the ecliptic at a rate of about 50 seconds a year. It drifts 30°, a whole zodiac constellation, in 2150 years. Then all star charts are out of date (Astronomers revise their precise star charts regularly.)

In astrology today, each sign of the zodiac bears the name of the constellation for which it was originally named but with which it no longer coincides due to precession of the equinoxes.

Refer to Figure 1.14. The present polestar is Polaris, and the vernal equinox is located in the constellation Pisces. (a) What was the polestar in the year 3000 B.C.? _____ (b) What will it be in the year A.D. 14,000? _____

Answer: (a) Thuban; (b) Vega.

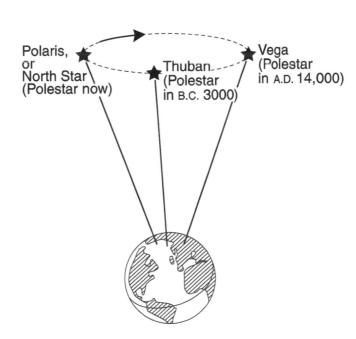

Figure 1.14. Precession. Earth's axis very slowly traces out a cone in space, so eventually the polestar changes.

SELF-TEST

This self-test is designed to show you whether or not you have mastered the material in Chapter 1. Answer each question to the best of your ability. Correct answers and review instructions are given at the end of the test.

1. For each of the following references used on a terrestrial globe, list the corresponding name on the celestial sphere:

   (a) Equator. _____

   (b) North Pole. _____

   (c) South Pole. _____

   (d) Latitude. _____

   (e) Longitude. _____

   (f) Greenwich, England. _____

2. Refer to Table 1.1. Which of the five brightest stars in the night sky are above the celestial equator, and which are below? _____
   _____

3. Refer to Table 1.1. Which of the five brightest nighttime stars never appear above the horizon at latitude 40° (about New York City)? _____
   _____

4. Match where you might be on Earth with the correct description of the stars:

   _____ (a) The stars seem to move along circles around sky parallel to your horizon.

   _____ (b) The stars rise at right angles to the horizon in the east and set at right angles to the horizon in the west.

   _____ (c) Vega practically crosses your zenith.

   _____ (d) Acrux is always above your horizon.

   _____ (e) Polaris appears about 30° above your horizon.

   (1) Antarctica (below 61°S).
   (2) Equator.
   (3) Jacksonville, Florida, U.S. (30°22′N).
   (4) North Pole.
   (5) Sacramento, California, U.S. (38°35′N).

5. Why do the stars appear to move along arcs in the sky during the night?

   _____

6. Why do some different constellations appear in the sky each season?

   _____

7. What is the zodiac? _____

   _____

8. Where on Earth would you have to be to have the Sun pass directly across
   your zenith at the time of the (a) vernal equinox? _____
   (b) summer solstice? _____ (c) winter solstice? _____

9. If a star rises at 8 P.M. tonight, at approximately what time will it rise a month
   from now?_____

10. Why is a solar day about 4 minutes longer than a sidereal day? _____

    _____

11. Arrange the following stars in order of decreasing brightness: Antares (mag-
    nitude 1); Canopus (magnitude –1); Polaris (magnitude 2); Vega (magnitude
    0). _____

12. Why will the polestar and the location of the vernal equinox on the celestial
    sphere be different thousands of years from now, causing your star maps
    finally to go out of date?_____

    _____

ANSWERS

Compare your answers to the questions on the self-test with the answers given below. If all of your answers are correct, you are ready to go on to the next chapter. If you missed any questions, review the sections indicated in parentheses following the answer. If you missed several questions, you should probably reread the entire chapter carefully.

1. (a) Celestial equator. (d) Declination.

   (b) North celestial pole. (e) Right ascension.

   (c) South celestial pole. (f) Vernal equinox.

   (Sections 1.1, 1.8, 1.9)

2. Above: Arcturus, Vega. Below: Sirius, Canopus, Rigil Kentaurus. (Sections 1.9, 1.10)

3. Canopus, Rigil Kentaurus. (Sections 1.10, 1.13, 1.14)

4. (a) 4; (b) 2; (c) 5; (d) 1; (e) 3. (Sections 1.10, 1.13 through 1.15)

5. Because of Earth's rotation. (Sections 1.1, 1.12, 1.14)

6. Because of Earth's revolution around the Sun. (Section 1.16)

7. A belt about 16° wide around the sky centered on the ecliptic, containing 12 constellations. (Section 1.17)

8. (a) Equator; (b) 23.5°N (Tropic of Cancer); (c) 23.5°S (Tropic of Capricorn). (Sections 1.19 through 1.21)

9. 6 P.M. (Section 1.16)

10. Because while Earth rotates on its axis, it also moves along in its orbit around the Sun. Earth must complete slightly more than one whole turn in space before the Sun reappears on your meridian. (Section 1.23)

11. Canopus, Vega, Antares, Polaris. (Section 1.7)

12. Because of the precession of Earth's axis. (Section 1.24)

2 LIGHT AND TELESCOPES

Curiosity is one of the permanent and certain characteristics of a vigorous mind.

Samuel Johnson (1709–1784)
The Rambler

Objectives

☆ Describe the wave nature of light, including how it is produced and how it travels.

☆ Name the major regions of the electromagnetic spectrum from the shortest wavelength to the longest.

☆ State the relationship between wavelength and frequency.

☆ State the relationship between the color of a star and its temperature.

☆ List the three windows (spectral regions) in Earth's atmosphere in order of their importance to observational astronomy.

☆ Explain how refracting and reflecting telescopes work.

☆ Define light-gathering power, resolving power, and magnification with respect to a telescope.

☆ State the two most important factors in telescope performance.

☆ State the purpose of a spectrograph.

☆ Explain how radio telescopes work, and list some interesting radio sources.

☆ Explain why infrared telescopes are located in very high, dry sites, and list some objects they observe.

☆ Explain why ultraviolet, X-ray, and gamma ray telescopes must operate above Earth's atmosphere, and list some objects they study.

Figure 2.1. Visualizing a light wave.

2.1 WHAT IS LIGHT?

Most of our information about the universe has been obtained through the analysis of starlight. To explain how starlight travels across trillions of kilometers of empty space to waiting telescopes, astronomers picture light as a form of wave motion.

A **wave** is a rising and falling disturbance that transports energy from a source to a receiver without the actual transfer of material. Wave motion is clearly observable in the ocean. During storms, crashing ocean waves vividly reveal the energy they carry.

A **light wave** is an electromagnetic disturbance consisting of rapidly varying electric and magnetic effects. Light waves transport energy from accelerating electric charges in stars (the source) to electric charges in the retina of your eye (the receiver) (Figure 2.1). You become aware of that energy when you see starlight.

What is a wave? _____

_____

Answer: A wave is a rising and falling disturbance that transports energy from a source to a receiver without the actual transfer of material.

2.2 WAVELENGTH

Light waves are distinguished by their lengths. The distance from any point on a wave to the next identical point, such as from crest to crest, is called the **wavelength** (Figure 2.2).

The human eye responds to waves that have extremely short wavelengths. Physicists and astronomers measure these waves in **nanometers**, **nm**, or the older **angstrom unit**, **Å**, after Swedish physicist Anders J. Ångstrom (1814–1874), who first measured wavelengths of sunlight. One nm is 10^{-9} m, and one angstrom is 0.10 nm.

The diameter of a human hair is about 50,000 nm (500,000 Å)!

Visible light has wavelengths of 400 nm to 700 nm. The varying wave-

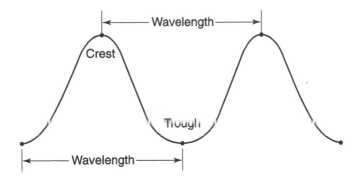

Figure 2.2. Wavelength measured from crest to crest or trough to trough.

lengths of visible light are perceived as different **colors**. The arrangement of the colors according to wavelength is called the **visible spectrum**.

Refer to Figure 2.3. Which color light has (a) the shortest wavelength? _____ (b) the longest wavelength? _____ (c) To which wavelength (color) is the eye most sensitive? _____

Answer: (a) Violet; (b) red; (c) 550 nm (yellow–green).

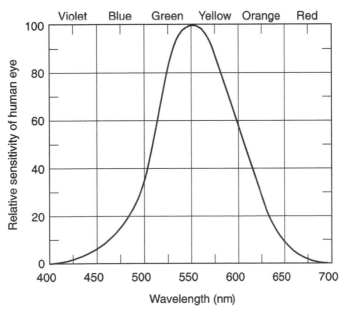

Figure 2.3. Relative sensitivity of the human eye to different colors and wavelengths of visible light.

2.3 THE ELECTROMAGNETIC SPECTRUM

Visible light is only one small part of all the electromagnetic radiation in space. Energy is also transmitted as gamma rays, X-rays, ultraviolet, infrared, submillimeter, microwave, and radio waves.

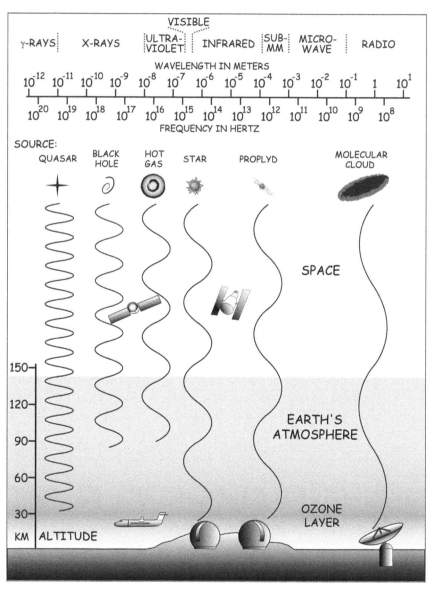

Figure 2.4. The electromagnetic spectrum includes all electromagnetic radiation from shortest, highest-frequency gamma rays to longest, lowest-frequency radio waves.

Because we make such different uses of them, these forms of radiation seem very different from one another. Doctors use gamma rays in cancer treatment and X-rays for medical diagnosis. Ultraviolet rays give you a suntan, and infrared rays warm you up. Radio waves are used for communication.

All of these forms of radiation are really the same basic kind of energy as visible light. They have different properties because they have different wavelengths. The shortest waves have the most energy, whereas the longest waves are the least energetic. The whole family of electromagnetic waves, arranged according to wavelength, is called the **electromagnetic spectrum**.

Electromagnetic waves of all wavelengths are important to astronomers because each type brings unique clues about its source.

Refer to Figure 2.4. List eight forms of electromagnetic radiation from the shortest waves (highest energy) to the longest waves (lowest energy). _____

_____

Answer: Gamma rays, X-rays, ultraviolet radiation, visible light, infrared radiation, submillimeter, microwave, radio waves.

2.4 RANGE OF WAVELENGTHS

What is the range of wavelengths included in the whole electromagnetic spectrum? _____

_____

_____

Answer: Wavelengths vary from less than a trillionth of a meter, 10^{-12} m, for the shortest gamma rays to longer than 10 m (33 feet) for the longest radio waves.

2.5 SPEED OF LIGHT

All kinds of electromagnetic waves move through empty space at the same speed—that is, at the speed of light. The speed of light in empty space, usually symbolized by the letter c, is practically 300,000 km/second (186,000 miles per second).

The speed of light in empty space has been called the "speed limit of the universe," because no known object can be accelerated to move faster. It is one of the most important and precisely measured numbers in astronomy (Appendix 2).

A **light-year (ly)** is the distance light travels through empty space in one year.

How many kilometers (miles) does 1 light-year represent? *Tips:* (1) distance = speed × time. (2) A year is equal to 3.156×10^7 seconds._____

Answer: Practically 9.5 trillion km (6 trillion miles).

Solution: Multiply 300,000 km/second × 3.156×10^7 second/year
(186,000 miles/second × 3.156×10^7 seconds/year)

2.6 WAVE FREQUENCY

Wave motion can be described in terms of frequency as well as wavelength. The **frequency** of a wave motion is the number of waves that pass by a fixed point in a given time, measured in **cycles per second (cps)**.

The human eye responds to different-color light waves that have very high frequencies. Visible light waves vary in frequency from 4.3×10^{14} cps for red to 7.5×10^{14} cps for violet, with the other colors in between.

For radio waves, one cycle per second is commonly called a **hertz (Hz)**, after the German physicist Heinrich Hertz (1857–1894), who first produced radio waves in a laboratory. An AM radio receives radio waves with frequencies of 550 to 1650 **KHz (kilohertz)**; 1 KHz is 1000 cycles per second. The FM band ranges from 88 to 108 **MHz (megahertz)**; 1 MHz is a million cycles per second.

Refer to the electromagnetic spectrum shown in Figure 2.4. Which waves have (a) a higher frequency than the visible light waves? _____
_____ (b) a lower frequency than the visible light waves? _____

Answer: (a) Higher frequency: gamma rays, X-rays, ultraviolet radiation. (b) Lower frequency: infrared radiation, submillimeter waves, microwaves, radio waves.

2.7 WAVELENGTH AND FREQUENCY

Can you deduce a general relationship between wavelength and frequency for these electromagnetic waves? _____
_____

Answer: The wavelength is inversely proportional to the frequency. The shorter waves have a relatively higher frequency, and the longer waves have a relatively lower frequency.

2.8 WAVE PROPAGATION

The relationship you have just found is an example of a formula that holds true for all kinds of wave motion:

$$\text{Speed of wave} = \text{Frequency} \times \text{Wavelength}$$

You can use this formula to calculate the frequency of any kind of electromagnetic wave in empty space if you know its wavelength (or the wavelength if you know the frequency). Explain why. *Tip*: Review Section 2.5._____

_____

_____

Answer: All electromagnetic waves have the same speed in empty space—that is, the speed of light, or about 300,000 km/second (186,000 miles per second).

2.9 WAVE EQUATION

Be sure you understand the relationship between speed (c), frequency (f), and wavelength (λ) for electromagnetic waves. The formula is:

$$c = f\lambda$$

Calculate the wavelength of a radio wave whose frequency is 100 KHz (100,000 cycles per second). _____

Answer: 3 km (1.86 miles).

Solution:

$$\text{Speed} = \text{Frequency} \times \text{Wavelength}$$

Thus,

$$\text{Wavelength} = \frac{\text{Speed}}{\text{Frequency}} = \frac{300,000 \text{ km/second}}{100,000 \text{ cycles/second}}$$

$$= \frac{186,000 \text{ miles/second}}{100,000 \text{ cycles/second}}$$

2.10 RADIATION LAWS

Stars, like other hot bodies, radiate electromagnetic energy of all different wavelengths. Energy due to temperature is called **thermal radiation**. The temperature of a star determines which wavelength is brightest.

Stars radiate energy practically as a **blackbody**, or theoretical perfect radiator. The intensity of radiation emitted over a range of wavelengths depends only on the blackbody's temperature. **Wien's law of radiation** states that the wavelength, λ_{max}, at which a blackbody emits the greatest amount of radiation is inversely proportional to its temperature (T). The formula is

$$\lambda_{max} = \frac{0.3}{T}$$

where λ_{max} is in centimeters and T is in kelvin (K). Thus the hotter a star, the shorter the wavelength at which it emits its maximum radiation.

Some stars are thousands of degrees hotter than others. You can judge how hot a star is by its color (wavelength). The hottest stars look blue-white (short wavelength), and the coolest stars look red (long wavelength).

Look in the sky for the examples cited in Table 2.1.

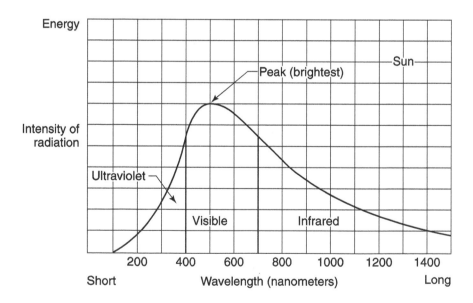

Figure 2.5. The Sun's thermal radiation spectrum. All blackbody radiation spectrums have the same shape. Hotter stars emit more energy at all wavelengths, and the peak shifts to shorter wavelengths.

TABLE 2.1 Four Hot and Cool Stars

Season	Star	Constellation	Color	Surface Temperature (K)
Summer	Vega	Lyra	Blue-white	10,000
Summer	Antares	Scorpius	Red	3,000
Winter	Sirius	Canis Major	Blue-white	10,000
Winter	Betelgeuse	Orion	Red	3,400

The **Stefan–Boltzmann radiation law** states that the total energy (E), emitted by a blackbody is proportional to the fourth power of its absolute temperature (T). Thus a star that is twice as hot as our Sun radiates 2^4, or 16, times more energy than the Sun.

A **radiation spectrum** shows how much energy a body radiates at different wavelengths, which wavelengths it radiates most intensely, and the total amount of energy it radiates at all wavelengths (indicated by the area under the curve).

Examine Figure 2.5. (a) The Sun radiates most intensely in the _____ wavelengths. (b) The total amount of energy that the Sun radiates as visible light is (more, less) _____ than the amount radiated outside the visible region.

Answer: (a) Visible; (b) less.

2.11 ASTRONOMICAL OBSERVATIONS

Today astronomers have tools to observe and analyze all forms of electromagnetic radiation from space. The main function of a telescope—whatever type of radiation is being detected—is to gather sufficient radiation for analysis.

Earth's atmosphere stops most radiation from space and permits only certain wavelengths to shine through to telescopes on the ground. Ground-based astronomers look out at the universe through two **atmospheric windows**, or spectral ranges within which air is largely transparent to radiation. These are the optical/visible light window, including some infrared, and the radio window.

An **astronomical observatory** is a place equipped for the observation of sky objects. For ground-based observations, astronomers prefer dark sites where the air is dry, thin, and steady, on mountaintops far from city lights and pollution (Figure 2.6).

Figure 2.6. Mauna Kea, a 4200-m-high (13,800-ft-high) site on the Island of Hawaii, U.S., hosts the world's largest group of optical, infrared, and submillimeter telescopes. Mauna Kea Visitor Information Station ▶http://www.ifa.hawaii.edu/info/vis◀ is at 3000 m (9200 ft).

What would you suggest to astronomers who want to observe the universe in the gamma ray, X-ray, and ultraviolet ranges? _____

_____

_____

Answer: Locate their instruments beyond Earth's atmosphere. Space-age technology makes space-based observations in these wavelength bands possible from rockets, spacecraft, or even Moon-based observing stations.

2.12 OPTICAL TELESCOPES

An **optical telescope** forms images of faint and distant stars. It can collect much more light from space than the human eye can. Optical telescopes are built in two basic designs—**refractors** and **reflectors**.

The heart of a telescope is its **objective**, a **main lens** (in refractors) or a **mirror** (in reflectors). Its function is to gather light from a sky object and focus this light to form an image. The ability of a telescope to collect light is called its **light-gathering power**.

Light-gathering power is proportional to the area of the collecting surface, or to the square of the **aperture** (clear diameter of the main lens or mirror). The **size** of a telescope, such as 150-mm or 8-m (6-inch or 26-foot), refers to the size of its aperture.

You can look at the image directly through an eyepiece, which is essentially a magnifying glass. Or you can photograph the image or record and process it electronically. Your eye lens size is about 5 mm (0.2 inch). A 150-mm (6-inch) telescope has an aperture over 30 times bigger than your eye lens. Its light-gathering power is 30^2, or 900 times greater than that of your eye. So a star appears over 900 times brighter with a 150-mm (6-inch) telescope than it does to your unaided eye.

Astronomers build giant telescopes to detect ever fainter and more distant objects.

All stars appear brighter with telescopes than they do to the eye alone. The extra starlight gathered by the telescope is concentrated into a single point. Using time exposure, a giant 10-m (400-inch) telescope can image very faint stars down to about magnitude 28, which is the same apparent brightness as a candle viewed from the Moon!

How much brighter would a star appear with the 10-m (33-foot) telescope than to your unaided eye? Explain. _____

_____

_____

Answer: Over 4 million times brighter. The 10-m (33-foot) telescope is over 2000 times bigger than your eye lens, so it gathers over 2000^2, or 4 million, times more light.

2.13 BINOCULARS

Binoculars are a practical first instrument for stargazing because they are easy to use and portable. A pair labeled 7×50 has an aperture of 50 mm. The $7\times$ specifies the magnification.

Why do binoculars and telescopes reveal many more sky objects than you can see with your unaided eye? _____

_____

_____

Answer: They can collect much more light than your eye can. (Light-gathering power is proportional to the square of the aperture.)

2.14 REFRACTING TELESCOPES

A refracting telescope has a main **objective lens** permanently mounted at the front end of a tube. Starlight enters this lens and is **refracted**, or bent, so that it forms an image near the back of the tube.

The distance from this lens to the image is its **focal length**. You may look at the image through a removable magnifying lens called the **ocular**, or **eyepiece**. The tube keeps out scattered light, dust, and moisture.

Italian astronomer Galileo Galilei (1564–1642) first pointed a refracting telescope skyward in 1609. The largest instrument he made was smaller than 50 mm (2 inches).

Today refracting telescopes range in size from a beginner's 60-mm (2.4-inch) to the largest ever built, the 1-m (40-inch) telescope at the Yerkes Observatory in Williams Bay, Wisconsin, U.S., which was completed in 1897.

Refer to Figure 2.7. Identify the refracting telescope's (a) objective lens; (b) the eyepiece; and (c) the focal length of the objective lens. State the purpose of (a) and (b). (a) _____

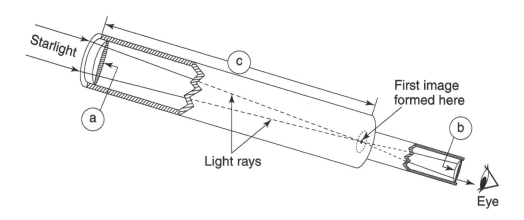

Figure 2.7. A refracting telescope with a long focal length objective lens and a short focal length eyepiece.

(b) _____

(c) _____

Answer: (a) Objective lens: to gather light and form an image. (b) Eyepiece: to magnify the image formed by the objective. (c) Focal length of objective lens.

2.15 REFLECTING TELESCOPES

A **reflecting telescope** has a highly polished curved-glass mirror, the **primary mirror**, mounted at the bottom of an open tube. When starlight shines on this mirror, it is reflected back up the tube to form an image at the **prime focus**.

You can record the image at the prime focus, or you can use additional mirrors to reflect the light to another spot. The **Newtonian telescope**, originated by British scientist Sir Isaac Newton in 1668, uses a small, flat mirror to reflect the light through the side of the tube to an eyepiece (Figure 2.8).

The **Cassegrain telescope** uses a small convex mirror, a **secondary mirror**, to reflect the light back through a hole cut in the primary mirror at the bottom end of the tube (Figure 2.9). It is more compact than a refractor or Newtonian reflector of the same aperture. The **Schmidt-Cassegrain** telescope combines an extremely short-focus spherical primary mirror at the back end of a sealed tube with a thin lens at the front.

Reflecting telescopes range in size from a beginner's 76-mm (3-inch) Newtonian reflector to the world's largest, the 10.4-m (34-foot) Gran Telescopio Canarias atop a peak on the Canary Island La Palma, Spain.

Figure 2.8. A Newtonian reflecting telescope with a primary mirror, a small diagonal secondary mirror, and an eyepiece.

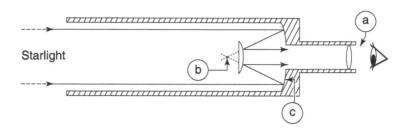

Figure 2.9. A Cassegrain reflecting telescope with a concave primary mirror, a small convex secondary mirror, and an eyepiece.

Refer to Figures 2.8 and 2.9. Identify the reflecting telescope's primary mirror; eyepiece; and prime focus. (a) _____ ; (b) _____ ; (c) _____

Answer: (a) Eyepiece; (b) prime focus; (c) primary mirror.

2.16 REFLECTORS VERSUS REFRACTORS

What is the essential difference between a reflecting telescope and a refracting telescope? Explain. _____

_____

Answer: The main optical part (objective). A reflecting telescope uses a mirror, whereas a refracting telescope uses a lens to collect and focus starlight.

2.17 f NUMBER

Telescopes are often described by both their aperture size and f number. The **f number** is the ratio of the focal length of the main lens or mirror to the aperture. These specifications are important because the brightness, size, and clarity of the image produced by a telescope depend on the aperture and focal length of its main lens or mirror.

For example, a "150-mm (6-inch), f/8 reflector" means the primary mirror is 150 mm (6 inches) in diameter and has a focal length of 1200 mm (8×150), or 48 inches (8×6).

What is the focal length of the 5-m (200-inch), f/3.3 mirror on Mount Palomar in California, U.S.?_____

Answer: 16.5 m (660 inches, or 55 feet).

2.18 IMAGES

All stars except our Sun are so far away that they appear as dots of light in a telescope. The Moon and planets appear as small disks. **Image size** is proportional to the focal length of the telescope's main lens or mirror.

For example, a mirror with a focal length of 2.5 m (100 inches) produces an image of the Moon that measures about 2.5 cm (1 inch) across. You know that the 5-m (200-inch), f/3.3 mirror has a focal length of 16.5 m (660 inches), which is over six times as long. Hence, it produces an image of the Moon that is about six times as big, or 15 cm (6 inches) across.

Lenses and mirrors form real images that are upside down. (A real image is formed by the actual convergence of light rays.) Since inverted images do not matter in astronomical work, and righting them would require additional light-absorbing optics, nothing is done to turn images upright in telescopes.

What determines the size of the image formed by a telescope? _____

Answer: The focal length of the main lens or mirror.

2.19 RESOLVING POWER

Even if a telescope were of perfect optical quality, it would not produce perfectly focused images because of the nature of light itself. A telescope's **resolving power** is its ability to produce sharp, detailed images under ideal observing conditions.

Resolving power depends directly on the size of the aperture and inversely on the wavelength of the incoming light. For the same light, a 150-mm (6-inch) telescope has twice the resolving power of a 75-mm (3-inch) telescope.

Starlight travels in straight lines through empty space, but when waves of starlight pass close to the edge of a lens or mirror, they spread out, in an effect called **diffraction**, and come to a focus at different spots. Because of diffraction, the image of a star formed by a lens or mirror appears under magnification as

Figure 2.10. Diffraction pattern (image of a star).

a tiny, blurred disk surrounded by faint rings, called a **diffraction pattern**, instead of as a single point of light (Figure 2.10). Diffraction limits resolving power.

If two stars are close together, their diffraction patterns may overlap so that they look like a single star. Features such as Moon craters and planet markings are also blurred by diffraction.

Resolving power determines the smallest angle between two stars for which separate, recognizable images are produced. The smallest resolvable angle for the human eye is about one minute of arc (1′), which is the size of an aspirin tablet seen at a distance of 35 m (110 feet).

Explain why what may look like a single star to the eye may resolve into two close neighbor stars in a telescope. _____

_____

Answer: Resolving power is proportional to aperture, and a telescope's aperture is much larger than the human eye's.

2.20 MAGNIFICATION

A telescope's **magnifying power** is the ratio of the apparent size of an object seen through the telescope to its size when seen by the eye alone. Telescopes magnify the angular diameter of objects. Thus the image appears to be closer than the object.

For example, to your eye the angular diameter of the full Moon is ½°, the same as an aspirin tablet held at arm's length (Figure 2.11). If the apparent size of the Moon increases 20 times, so that it looks 10° in diameter when you view it through your telescope, then the magnifying power is 20, written 20×.

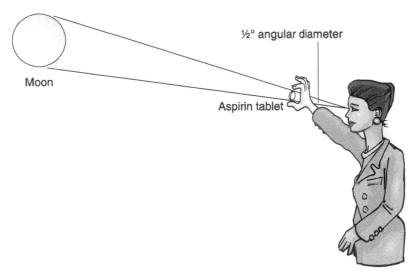

Figure 2.11. Angular diameter.

The value of the magnifying power of a telescope depends on the eyepiece you use. You can figure:

$$\text{Magnifying power} = \frac{\text{Focal length of telescope}}{\text{Focal length of eyepiece}}$$

A telescope usually comes with several eyepieces of different focal lengths so you can vary its magnifying power for viewing different objects.

(a) What is the magnifying power of a 150-mm (6-inch), f/8 telescope when an eyepiece of 12.5-mm (½-inch) focal length is used? _____ (b) How could you increase the magnifying power of this telescope? _____

Answer: (a) 96×.

Solution:

$$\text{Magnifying power} = \frac{\text{Focal length of telescope}}{\text{Focal length of eyepiece}}$$

$$= \frac{1200 \text{ mm}}{12.5 \text{ mm}} = \frac{48 \text{ inches}}{\frac{1}{2} \text{ inch}}$$

(b) Use an eyepiece of a shorter focal length.

2.21 MAXIMUM USEFUL MAGNIFICATION

It is a mistake to exaggerate the importance of magnifying power when you buy a telescope. You cannot increase the useful magnifying power indefinitely by changing eyepieces.

Starlight must pass through Earth's atmosphere to reach waiting telescopes on the ground. Disturbances in the air cause blurry images. **Seeing** refers to atmospheric conditions that affect the sharpness of a telescope's image. If the air is quiet, then the seeing is good, and stars shine with a steady light. If the air is turbulent, then the seeing is bad, and stars twinkle madly.

The **practical limit of useful magnification** for any telescope is about two times its aperture in millimeters (50 times its aperture in inches). Higher power will just magnify any blurring in the image due to diffraction or bad seeing. It cannot reveal any finer details.

A telescope in space escapes interference from Earth's atmosphere, so it can see farther and image sharper than a telescope on the ground (Figure 2.12). Astronomers operate space observatories by remote control from the ground. Astronauts can maintain, repair, and upgrade a space telescope in orbit around Earth or bring it home for a major overhaul.

The U.S./European Hubble Space Telescope (HST) (1990–) sends us amazing images of planets, stars, and other objects out to the farthest galaxies. ▶hubblesite.org◀ HST orbits a 2.4-m (94-inch) main mirror and eight instruments for visible, ultraviolet, and infrared light observations (Figure 2.13).

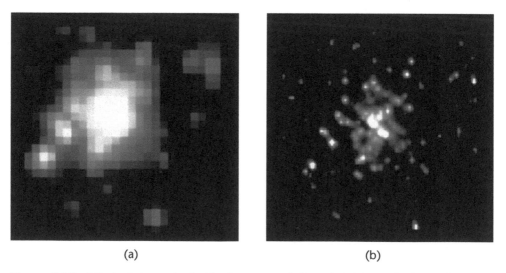

(a)   (b)

Figure 2.12. Effect of atmospheric blurring on resolution. A tightly packed star cluster in the 30 Doradus Nebula located 163,000 light-years away, imaged with (a) a ground-based telescope at the European Southern Observatory, Chile; and (b) Hubble Space Telescope.

Figure 2.13. Hubble Space Telescope data path. HST is a 13-m (43-foot) serviceable space observatory that orbits 600 km (380 miles) above Earth. Instruments are powered by two solar arrays. Communication is via the Tracking and Data Relay Satellite System (TDRSS).

What is the practical limit of useful magnification for a 150-mm (6-inch) telescope? _____

Answer: 300×.

2.22 TELESCOPE ABERRATIONS

Aberrations are imperfections in the image produced by an optical system.

Chromatic aberration is a lens defect. Starlight consists of all the colors of the spectrum. When starlight passes through a lens, the lens focuses different colors (wavelengths) at slightly different distances. This variation blurs the star image with spurious colors (Figure 2.10). An **achromatic lens**, a combination of two or more lenses made of different kinds of glass, counters this defect.

A properly curved mirror reflects all the colors of starlight to a focus at the same point. The image formed by a reflector has no blurred colors.

Spherical aberration is a mirror defect that blurs a star image. It is a

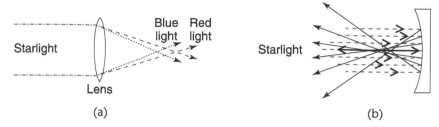

Figure 2.14. (a) Chromatic aberration. A lens bends blue (shorter) light waves the most and brings them to a focus closer to the lens than red (longer) light waves. (b) Spherical aberration. An improperly curved mirror does not reflect light waves to a single focus.

defect of spherical surfaces, hence the name. Parts of the mirror at different distances from the optical axis reflect starlight to slightly different focal points (Figure 2.14).

A **parabolic mirror** avoids this defect. Its paraboloid shape is curved less at the edges than at the center, and so it properly reflects starlight to a single focus. A **catadioptric telescope**, a refractor-reflector combination, has a correcting lens or plate at the upper end of the tube to correct the aberrations of a primary mirror with a spherical shape.

Why should you have the best-quality optical parts in your telescope? _____

Answer: To avoid image aberrations.

2.23 TELESCOPE DESIGN AND SELECTION

You probably wonder which type of telescope is better—a refractor or a reflector. The answer depends on the application involved since each type has advantages and disadvantages over the other.

Small telescopes for hobbyists can be of either design. Refractors, with their sealed tubes, are rugged and require less maintenance. But reflectors offer greater aperture for the price and are easier to make at home. The **Dobsonian telescope**, a Newtonian reflector on a simple mount, is popular because it is easy to use and cheapest for a given size. Although more expensive per unit of aperture, catadioptric telescopes such as the Schmidt-Cassegrains and **Maksutov-Cassegrains** are the most compact and portable.

Whatever design you choose, the stability of your small telescope **mount** is essential. Nothing will kill your enthusiasm for stargazing faster than a poor-quality telescope with a shaky mount that provides blurry, wiggling images.

Large refractors are used where image quality and resolution are most important, as for viewing surface details of the Moon and planets or for observing double star systems.

Giant reflectors are used where aperture is most important, as to probe the faintest, most distant objects. They are easier to build and are more cost effective than refractors. **Folded optics** reduce the physical length of huge reflectors, so they can be housed inside smaller domes than refractors. The primary mirror is supported from behind so it does not sag under gravity as large lenses do.

Astronomers design ever-larger telescopes and new observing techniques to increase light collection and improve resolving power (Figure 2.15).

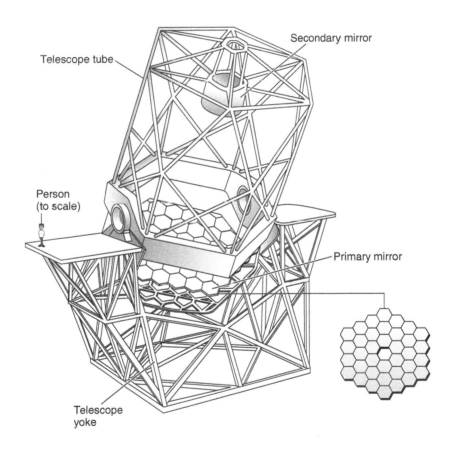

Figure 2.15. The Keck Telescope in Hawaii, U.S., uses a segmented-mirror design for optical and infrared research. Computer controls precisely align 36 hexagonal mirrors, each about 2 m (6 feet) across and 7.5 cm (3 inches) thick, to form one surface and function as a huge 10-m (33-foot) mirror. Keck I and twin Keck II can operate independently or together. ▶www.keckobservatory.org◀

The newest telescopes have lighter-weight **monolithic mirrors**, cast as a single piece, or **segmented mirrors**, mosaics of individual mirrors (Figure 2.15) that are used both independently or in combination, with a computer control system. **Multiple telescopes**, more than one primary mirror, send light collected by all primaries to a central focus where it is combined to image as a single gigantic virtual mirror.

The largest operating telescope in the southern hemisphere is the European Southern Observatory's 16-m Very Large Telescope (VLT). ▶www.eso.org.paranal◀ The VLT has a multiple-mirror design utilizing four linked 8.2-m telescopes that were named in the local Mapuche language.

Most of the world's biggest telescopes have interesting visitor websites ▶http://astro.nineplanets.org/bigeyes.html◀ and self-guided tours for the public (Table 2.2).

TABLE 2.2 Major Optical Telescopes in the World

Project	Mirror Size (m)	Observatory Location	Description
European Extremely Large Telescope (E-ELT)*	39.0	European Southern Cerro Armazones, Chile	Giant primary of 798 hexagonal 1.4 m mirror segments; optical to near-infrared wavelengths
Giant Magellan Telescope*	21.4	Las Campanas La Serena, Chile	Seven 8.4-m primary mirrors—six off-axis surrounding the seventh central on axis; resolving power of a 24.5-m primary mirror
Thirty Meter Telescope*	30.0	TMT Corporation Mauna Kea, HI	Segmented mirror of 492 components; ultraviolet to mid-infrared wavelengths
Gran Telescopio Canarios	10.4	Roque de los Muchachos Canary Islands, Spain	Segmented mirror of 36 hexagonal components based on Keck design
Keck I Keck II	10.0	W. W. Keck Mauna Kea, HI	Two mirrors of 36 segments each; 85 meters apart; usable as optical interferometer (Figure 2.15)
South African Large Telescope (SALT)	~10	S. African Astronomical Sutherland, S.A.	Multinational, hexagonal mirror array, fixed elevation, based on Hobby-Eberly design
Hobby-Eberly	9.2	MacDonald Mt. Fowlkes, TX	Spherical segmented mirror, fixed elevation mounting; spectroscopy only
Large Binocular Telescope	8.4	LBT Corporation Mt. Graham, AZ	Two 8.4-m mirrors on one mount give light gathering of an 11.8-m and resolution of a 22.8-m mirror
Subaru Telescope	8.3	Japan Nat'l Astronomy Mauna Kea, HI	Lightweight (22.8t) meniscus primary mirror 20-cm thick; active support
Very Large Telescope	8.2	European Southern Cerro Paranal, Chile	Four separate 8.2-m telescopes: Antu (Sun), Kueyen (Moon), Melipal (Southern Cross), Yepun (Venus) or combined as a 16.4-m aperture
Gillet Gemini North Gemini South	8.0	Gemini[a] Mauna Kea, HI Cerro Pachon, Chile	Multinational optical/infrared twins offer unobstructed coverage of both northern and southern skies

*Under construction.

[a]Facilities of the National Optical Astronomy Observatories (NOAO), headquartered in Tucson, AZ.

Large telescope performance is dramatically improved by new techniques. **Adaptive optics** adjust flexible mirrors to correct for rapid, hundredths-of-a-second distortions due to turbulence in Earth's atmosphere. **Active optics** correct for minute- or hour-long mirror-shape distortions due to gravity, temperature drifts, and wind.

What are the main advantages of an optical telescope over the unaided eye?

_____

_____

Answer: Superior light-gathering power and resolution. A telescope can also be equipped to record light over a long period of time.

2.24 TELESCOPE ENHANCEMENTS

Research time is in great demand, so astronomers do not sit at giant telescopes and simply stargaze. Instead, observers usually look at a computer display! Starlight, directly or after passing through electronic imaging systems, is recorded for exhaustive study later by many scientists and for obtaining pictures. Powerful **computers** are vital to the acquisition, archiving, processing, and analysis of astronomical data today.

A **charge-coupled device (CCD)** is a popular electronic detector. The CCD is a silicon chip of tiny, light-sensitive elements that turns starlight into electric pulses for computers and advanced image processing and display equipment. CCDs are much more sensitive to light than photographic film, and they can record bright and faint objects simultaneously.

Often an instrument called a **spectrograph** is attached to the telescope. Starlight is not a single color but rather a mixture of colors, or wavelengths (Figure 2.16). Astronomers deduce much information about stars from these separate wavelengths, as you will see in Chapter 3.

Figure 2.16. You can produce a spectrum from sunlight (starlight). Place a mirror in a pan of water so that it is under the water and leaning against the side of the pan. Position the pan in bright sunlight so that the Sun shines on the mirror. Move the mirror slightly until you see a spectrum on the ceiling or wall.

A **spectroscope** separates starlight into its component wavelengths for viewing. Starlight enters the spectroscope through a narrow slit and goes through a collimating lens, which produces a beam of parallel rays of light. A prism or grating disperses this light into its separate colors (wavelengths). This spectrum is recorded in a spectrograph.

What is the purpose of a spectrograph?_____

_____

Answer: To separate and record the individual wavelengths in a beam of light.

2.25 RADIO ASTRONOMY

New kinds of telescopes allow today's astronomers to "look" farther into space and "see" more fascinating sights than at any time in the past.

Most **radio telescopes** use a curved "dish" **antenna**, which corresponds to the main mirror in an optical reflector, to collect and focus radio waves from space. This antenna must be very big to collect long radio waves and produce clear images (Figure 2.17).

Figure 2.17. A radio telescope at the Karl Jansky Very Large Array.

You cannot see, hear, or photograph these radio waves directly. Instead, they are redirected to a tuned radio receiver that amplifies, detects, and records their electronic image. Computers may display radio images digitized, as a contour map that shows the strength of the radio source (Figure 6.19b) or as a **radiograph** (Figure 6.18), which is a false color picture that shows how the radio source in space would "look" to a person with "radio vision."

Radio astronomy began in 1931 when U.S. engineer Karl G. Jansky (1905–1950) discovered radio waves coming from the Milky Way. Since then radio waves have been received from diverse sources, including our Sun, planets, cold interstellar gas, pulsars, distant galaxies, and quasars.

The largest single radio antenna ever built is a 305-m (1000-foot) dish with an 8-hectare (20-acre) reflecting area. It is fixed in a valley at the Arecibo Observatory, Puerto Rico. ▶www.naic.edu◀

The Robert C. Byrd Green Bank Telescope (GBT), at the National Radio Astronomy Observatory (NRAO) in West Virginia, U.S., ▶www.nrao.edu◀ is the world's most powerful, accurate, and sensitive fully steerable single antenna. Its innovative 100 × 110-m (330-foot) dish is specially shaped to direct radio waves to the side, where a receiver collects the signals without blocking the dish.

Identify the antenna and prime focus of the radio telescope shown in Figure 2.17. (a) _____ ; (b) _____

Answer: (a) Prime focus; (b) antenna.

2.26 RADIO TELESCOPES

Radio telescopes have several advantages. They let us "see" many celestial objects that emit powerful radio waves but little visible light. They let us "see" radio sources behind interstellar dust clouds in our Milky Way Galaxy that blot out visible stars (because radio waves pass through these clouds). Our atmosphere does not stop or scatter radio waves, so radio telescopes can be used in cloudy weather and during the daytime.

As with optical telescopes, more and clearer data are produced by ever-larger collectors. **Aperture synthesis** is a cost-effective way to get the performance of a single giant telescope from smaller ones. An **interferometer** combines beams of light from two or more telescopes to simulate one very large aperture whose resolving power is set by the separation of the smaller ones.

The Karl G. Jansky Very Large Array (VLA) is the world's largest aperture synthesis facility. It has 27 movable 25-m (82-foot) antennas located at a 2100-m (7000-foot) high National Radio Astronomy Observatory site in New Mexico, U.S. These can be used in different configurations to act as a fully steerable radio dish 36 km (22 miles) in diameter. Computers control the antennas, analyze and display observed data, and produce top-quality, detailed images.

Very Long Baseline Interferometry (VLBI) gives the best resolution by stationing antennas continents apart. Each station has receiving, transmitting, data handling, and interstation communication equipment. Data recorded from coordinated observations of a specific radio source is correlated by computer to simulate one colossal dish.

The U.S. Deep Space Network (DSN) ▶http://deepspace.jpl.nasa.gov/dsn◀ has three 70-m (230-foot) radio telescopes set in California, U.S., Spain, and Australia. Stations are used for VLBI observations 24/7/365 and for constant contact with spacecraft as Earth rotates. The control center is at the NASA Jet Propulsion Laboratory in California.

The Very Long Baseline Array (VLBA) maps the most distant radio sources and finest details (Figure 2.18). It has 10 automated 25-m (82-foot)

Figure 2.18. The Very Long Baseline Array (VLBA) ▶www.vlba.nrao.edu◀ has 10 antennas at stations located across 5000 miles. The fine detail that the VLBA can "see" is like being able to stand in New York and read a newspaper in Los Angeles.

radio telescopes set across the U.S. from Hawaii to St. Croix, Virgin Islands, with an operations center in New Mexico. Astronomers monitor their research while VLBA operators remotely control antennas and check equipment over the Internet. Supercomputer processing of the recorded data from all 10 antennas subsequently can synthesize a single radio telescope 8000 km (5000 miles) in diameter.

Resolving power is maximized by using the VLBA with Earth-orbiting radio telescopes.

List at least three advantages of a radio telescope. (1) _____

_____ ;

(2) _____ ;

(3) _____

Answer:

(1) Reveals radio sources—objects that shine in the radio band of wavelengths.

(2) Shows radio sources behind interstellar dust clouds in parts of the Milky Way Galaxy that are hidden from optical viewing.

(3) Works in cloudy weather and during daytime.

(4) Shows radio sources that are located beyond our power of optical viewing.

2.27 INFRARED ASTRONOMY

Infrared astronomy studies incoming radiation with wavelengths beyond visible red to radio.

Infrared telescopes were first built in the 1960s. They are basically optical reflectors with a special heat detector at the prime focus. Detectors are shielded and cooled to about 2 K to ensure that they register infrared rays from space, rather than stray heat from people, equipment, and observatory walls.

Water vapor and carbon dioxide in the air strongly absorb infrared rays. Large infrared telescopes are located on very high mountaintops where the air overhead is thinnest and driest. Smaller telescopes are lofted in airplanes, balloons, rockets, and spacecraft.

U.S./German Stratospheric Observatory for Infrared Astronomy (SOFIA) ▶http://www.sofia.usra.edu/◀ is an airplane modified to fly a 2.5-m reflecting telescope above 12-km (40,000 feet). U.S. Spitzer Space Telescope (2003–) ▶www.spitzer.caltech.edu◀ orbits an 85-cm telescope.

Infrared telescopes image invisible sources that are relatively cool or obscured because infrared rays pass through interstellar clouds of gas and dust that block shorter visible rays. You can see false color images of cool stars and galaxies, regions of star and planet formation in giant molecular clouds, comets, and galaxy centers at NASA's Infrared Processing and Analysis Center (IPAC). ▶www.ipac.caltech.edu◀

What is the main advantage of infrared telescopes?

Answer: They reveal relatively cool objects that may not be visible.

2.28 ULTRAVIOLET, X-RAY, AND GAMMA RAY ASTRONOMY

Since the 1960s, **ultraviolet**, **X-ray**, and **gamma ray telescopes** with suitable detectors have been sent above Earth's obscuring air in orbiting spacecraft.

Solar arrays collect and convert sunlight to electricity for instruments and directional control. Insulation protects instruments from the extreme heat and cold, low pressure, and energetic particles and radiation in space. Star trackers and gyroscopes orient space observatories and point them to sky objects on command.

High energy telescopes collect and focus incoming radiation. Detectors record its intensity, energy, duration, and direction of origin. Radio antennas receive commands from mission ground control and transmit data to the ground.

The data are processed and recorded by computer for analysis. They are displayed digitally or as graphs of intensity over time or an energy range to reveal how the source is producing its rays, how bright it is, how long it remains at that brightness, and what kind of object it is. Data can be manipulated to generate spectacular **false color images**, in which colors are used to show features of invisible objects (not colors you would actually see).

Ultraviolet observations of the Sun, hot stars, stellar atmospheres, interstellar clouds, a hot gas galactic halo, and extragalactic sources abound. The U.S. robot Galaxy Evolution Explorer (GALEX) (2003–) probes the faintest and most distant sources ever. ▶http://galex.caltech.edu◀

X-rays and gamma rays shoot right through ordinary mirrors and lenses, so we use alternate ways to image the most energetic objects and violent events in the universe. U.S. robot Chandra X-ray Observatory (1999–) ▶http://chandra.harvard.edu◀ has nested barrel-shaped mirrors. Incident X-rays that strike them at grazing angles bounce off to a focus and form an image. Gamma ray detectors show the telling spray that appears after they are absorbed or collimated to a collision in a high-density medium (Figure 2.19).

X-ray and gamma ray telescopes reveal sudden, intense bursts of radiation (**bursters**), possible black holes, active galaxies, and distant quasars.

Figure 2.19. Fermi Gamma-ray Large Area Space Telescope (2008–). Large Area Telescope (LAT) tracker and calorimeter measure direction and energy of incoming gamma rays; outer detector bans other particles. A complementary Glast Burst Monitor (GBM) detects X-rays and less energetic gamma rays. ▶fermi.gsfc.nasa.gov◀

What is particularly interesting about new observations by ultraviolet, X-ray, and gamma ray telescopes? _____

_____

_____

Answer: Incoming ultraviolet, X-rays, and gamma rays have much more energy than visible light. They must be generated in extraordinarily energetic processes not yet fully comprehended.

SELF-TEST

This self-test is designed to show you whether or not you have mastered the material in Chapter 2. Answer each question to the best of your ability. Correct answers and review instructions are given at the end of the test.

1. Explain why looking at stars is a way of seeing how the universe looked many years ago. _____

    _____

    _____

    _____

    _____

2. (a) List the major regions of the electronic spectrum from shortest wavelength (highest energy) to longest wavelength (lowest energy). _____

    _____

    _____

    _____

    (b) State what all electromagnetic waves have in common. _____

    _____

3. Write the general formula that relates the wavelength and frequency of a wave. _____

4. Suppose you observe a bluish star and a reddish star. State which is hotter, and explain how you know. _____

    _____

    _____

5. List the two windows (spectral ranges) in Earth's atmosphere for observational astronomy. _____

6. What are the two main parts of a telescope used for stargazing, and what is the function of each?_____

   _____

7. What are the two main advantages of giant telescopes for research?

   _____

Two Telescopes

Characteristic	Type of Telescope	
	Reflector (1)	Refractor (2)
Diameter of main lens or mirror	2 m	1 m
Focal length of objective	7.6 m	14.6 m
Focal length of eyepiece	5 cm	1 cm

8. Which telescope described in the chart above (1 or 2) has:

   (a) greater light-gathering power?

   (b) greater resolving power?

   (c) greater magnification?

9. What two factors are most important in telescope performance?_____

   _____

10. What is the purpose of a spectrograph? _____

    _____

11. List three advantages of a radio telescope. _____

    _____

    _____

12. What is the advantage of sending telescopes up in spacecraft?_____

    _____

    _____

    _____

13. Match an appropriate innovative tool to the observations.

_____ (a) Faintest and most distant radio sources.

_____ (b) Very hot stars and gas.

_____ (c) Visible and relatively cool sources.

_____ (d) X-ray sources.

(1) Chandra X-ray Observatory.

(2) Galaxy Evolution Explorer (GALEX).

(3) Keck Telescope.

(4) Very Long Baseline Array (VLBA).

ANSWERS

Compare your answers to the questions on the self-test with the answers given below. If all of your answers are correct, you are ready to go on to the next chapter. If you missed any questions, review the sections indicated in parentheses following the answer. If you missed several questions, you should probably reread the entire chapter carefully.

1. Starlight is radiated by electric charges in stars. Light waves transport energy from stars to electric charges in our eyes. Light waves travel incredibly fast—about 300,000 km (186,000 miles) per second. But trillions of miles separate the stars from Earth, and the journey takes many years. Thus we see the stars as they were many years ago when the starlight began its journey to Earth. (Sections 2.1, 2.5)

2. (a) Gamma rays, X-rays, ultraviolet radiation, visible light, infrared radiation, submillimeter, microwave, radio waves. (b) All electromagnetic waves travel through empty space at the same speed, the speed of light—about 300,000 km (186,000 miles) per second. (Sections 2.3, 2.5, 2.8)

3.
$$c = f\lambda \text{ or Wavelength} = \frac{\text{Speed of wave}}{\text{Frequency}}$$

(Sections 2.2, 2.5, 2.6, 2.8, 2.9)

4. The bluish star is hotter. The shorter the wavelength at which a star emits its maximum light, the hotter the star, according to Wien's law of radiation. Blue light has a shorter wavelength than red light. (Sections 2.2, 2.10)

5. Optical (visible light) including infrared; radio. (Section 2.11)

6. (1) Main mirror or lens (objective): To gather light and form an image. (2) Eyepiece: To magnify the image formed by the main mirror or lens. (Sections 2.12, 2.14, 2.15)

7. Superior light-gathering power and resolving power. (Sections 2.12, 2.19, 2.23)

8. (a) 1; (b) 1; (c) 2. (Sections 2.12, 2.19, 2.20)

9. Size and quality of main mirror or lens. (A stable mount is essential.) (Sections 2.12, 2.17 through 2.23)

10. To separate and record the individual wavelengths in a beam of light. (Section 2.24)

11. Reveals radio sources; shows radio sources that are hidden from sight behind interstellar dust clouds in the Milky Way Galaxy; works in cloudy weather and daytime; shows radio sources that are located beyond our power of optical viewing.   (Sections 2.25, 2.26)

12. Spacecraft take the telescopes beyond Earth's obscuring atmosphere, and it is possible to observe gamma ray, X-ray, and ultraviolet sources that cannot be observed on the ground. There is no atmospheric blurring or radio interference, so a space telescope can work at its practical limit of resolving power.   (Sections 2.11, 2.21, 2.26, 2.27, 2.28)

13. (a) 4; (b) 2; (c) 3; (d) 1.   (Sections 2.23, 2.26, 2.28)

3

THE STARS

Look at the stars! look, look up at the skies!
O look at all the fire-folk sitting in the air!
Gerard Manley Hopkins (1844–1899)
"The Starlight Night"

Objectives

☆ Describe the method and range of the parallax technique for determining the distances to stars.

☆ Describe three types of spectra: emission, absorption, and continuous spectra.

☆ Explain why emission and absorption spectra are unique for each element.

☆ Give a general description of stellar spectra, and explain how they are divided into spectral classes.

☆ Explain how a star's chemical composition, surface temperature, and radial velocity are determined from its spectrum.

☆ List several other kinds of information that are obtained from stellar spectra.

☆ Explain how a star's proper motion and space velocity are determined.

☆ Explain the difference between apparent brightness and luminosity.

☆ Explain the relationship between apparent magnitude, absolute magnitude, and distance.

☆ Describe the H–R diagram and explain the relationship of a star's mass to its luminosity and temperature.

☆ Compare red giants and white dwarfs with our Sun in terms of mass, diameter, and density.

☆ Define four types of binary star systems.

3.1 DISTANCES TO NEARBY STARS

Stars are really trillions of kilometers beyond our atmosphere. The difficult problem of ascertaining the actual positions and motions of stars, called **astrometry,** has challenged astronomers for centuries.

The method of **parallax** is used in measuring the distances to nearby stars. The position of a star is carefully determined relative to other stars. Six months later, when Earth's revolution has carried telescopes halfway around the Sun, the star's position is measured again.

Nearby stars appear to shift back and forth relative to more distant stars as Earth revolves around the Sun. The apparent change in a star's position observed when the star is sighted from opposite sides of Earth's orbit is called **stellar parallax.** The distance to the star is calculated from its **parallax angle**, which is one half of the apparent change in the star's angular position (Figure 3.1).

Stellar parallaxes are very small and are measured in **seconds of arc** ("), where 1" = 1/3600°. An aspirin tablet would appear to have a diameter of 1" if it were viewed from a distance of about 2 km (a mile)! The parallaxes of even the nearest stars are less than 1"(Appendix 5).

One **parsec (pc)**, is the distance to an imaginary star whose *par*allax is 1 *sec*ond of arc (1"). One parsec equals about 31 trillion km (19 trillion miles), or 3.26 light-years.

To calculate the distance to any star from its measured parallax, use the formula:

$$\text{Star's distance (in pc)} = \frac{1}{\text{parallax (")}}$$

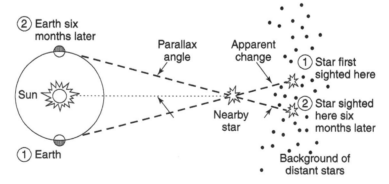

Figure 3.1. Stellar parallax. A nearby star that is sighted from opposite sides of Earth's orbit appears to shift its position from 1 to 2 against a background of distant stars. (The actual parallax angle is extremely tiny.)

Parallax angles are hard to measure through Earth's blurring air, and they decrease with the distance of a star. The best measurements, to about 0".002 or 500 pc, are from ESA High Precision Parallax Collecting Satellite, or Hipparcos, (1989–1993). The resulting Hipparcos Star Catalog lists measured positions, parallaxes, and motions of 118,000 stars precisely and the subsequent Tycho-2 Catalog lists data for another 2 million stars less exactly.

Now ESA spacecraft GAIA (2013–) aims to make the largest, most precise three-dimensional map of our Milky Way Galaxy from unprecedented measurements to microarcsecond accuracies of position and motion for about a billion stars – nearly 1 percent of all the stars in our Galaxy.

Would you like to know what "close" means for a star? Refer to Figure 3.2. If the measured parallax for Alpha Centauri is 0".74, then its distance from Earth is 1.35 pc, or 4.4 light-years, which is about 40 trillion km (25 trillion miles). (Alpha Centauri is actually a double star and, with faint Proxima Centauri—the night star closest to us—a member of a triple star system.)

Other indirect methods must be used to determine the distances to the great majority of stars farther away.

If the measured parallax of Sirius is 0".38, what is its distance from Earth in (a) parsecs? _____ (b) light-years? _____ (c) kilometers or miles (approximate)? _____

Answer: (a) 2.6 pc; (b) 8.6 ly; (c) 81 trillion km or 50 trillion miles.

Solution:

(a) $\dfrac{1}{0".38}$;   (b) (2.6 pc) x 3.26 $\dfrac{ly}{pc}$;

(c) (2.6 pc) x 31 trillion $\dfrac{km}{pc}$

(2.6 pc) x 19 trillion $\dfrac{miles}{pc}$

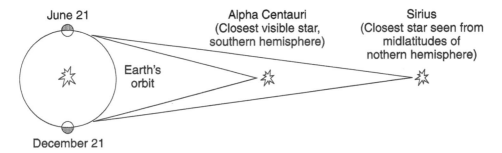

Figure 3.2. Using the parallax method to determine the distances to our closest bright neighbor stars.

3.2 TYPES OF SPECTRA

Despite the vast distances that separate us from the stars, we know a lot about them. Astronomers can extract an amazing amount of information from starlight.

Remember that starlight is composed of many different wavelengths. When starlight is separated into its component wavelengths, the resulting spectrum holds many clues about the stars. **Spectroscopy** is the analysis of *spectra* (or *spectrums*). Spectra are of three basic types, each produced under different physical conditions.

Describe the appearance of each type of spectrum illustrated in Figure 3.3.

(a) _____

(b) _____

(c) _____

Answer:

(a) **Continuous spectrum**: a continuous array of all the rainbow colors.

(b) **Emission**, or **bright-line, spectrum**: a pattern of bright-colored lines of different wavelengths.

(c) **Absorption**, or **dark-line, spectrum**: a pattern of dark lines across a continuous spectrum.

Note: Modern astronomers work with spectra as graphs of intensity versus wavelength (Figures 3.8, 3.9, 6.22b).

Figure 3.3. The three basic types of spectra as viewed through a spectroscope.

3.3 SPECTRAL LINES

Atoms are responsible for each type of spectrum. An **atom** is the smallest particle of a chemical element.

More than 100 chemical elements have been identified (Appendix 4). Each element has its own particular kind of atom, first simply described by Danish physicist Niels Bohr (1885–1962).

In the **Bohr atom model**, each element's atoms have a nucleus with a unique number of positively charged protons, circled by the same number of electrons bearing a corresponding negative charge. Atoms are normally electrically neutral.

The electrons are confined to a set of allowed orbits of definite radius. An electron in a particular orbit has a definite **binding energy**, the energy required to remove it from the atom. Each element has its own unique set of allowed electron orbits, or **energy levels**.

An undisturbed atom, in the **ground state**, has the least possible energy.

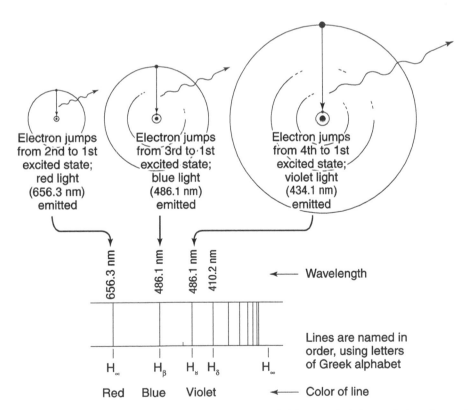

Figure 3.4. Origin of the unique set of bright red, blue, and violet emission lines of hydrogen. (Ground state not shown.)

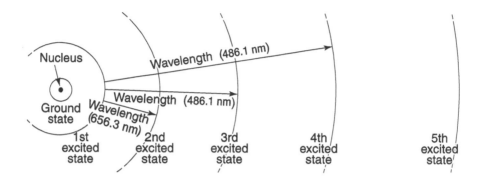

Figure 3.5. Origin of the dark absorption lines corresponding to the bright red, blue, and violet emission lines of hydrogen.

If the right energy is supplied, an electron will jump to a higher energy level. Then the atom is in an unstable **excited state**. When the electron falls back down, the atom radiates that energy in the form of a pellet of light called a **photon**.

If an atom absorbs enough energy, one or more of its electrons can be removed completely. The atom, which is left with an electric charge (positive), is then called an **ion**.

Bright-colored **emission lines** are produced when electrons jump from higher energy levels back down to lower energy levels. The wavelength of the light emitted is inversely proportional to the energy difference between the energy levels. Since each kind of neutral or ionized atom has its own unique set of energy levels, each chemical element has its own unique set of bright-colored emission lines (Figure 3.4).

Corresponding unique **dark absorption lines** are produced when an atom of a chemical element absorbs light and the electrons jump out to higher energy levels (Figure 3.5).

Thus, an emission spectrum or a corresponding **absorption spectrum** gives positive identification of the chemical element that produced it.

Why do atoms emit light of different colors (specific wavelengths)? _____

_____

_____

Answer: Each color (wavelength) corresponds to an electron jumping down from a particular higher energy level to a particular lower energy level.

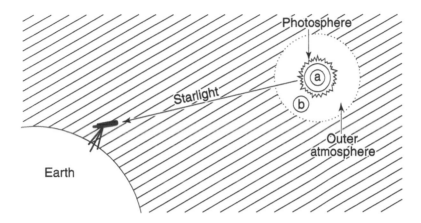

Figure 3.6. The starlight we observe comes from a star's photosphere and passes through its outer atmosphere before shining into space.

3.4 SPECTRA OF STARS

Stellar spectra, or the spectrums of stars, are absorption spectrums. Photographs show dark lines crossing a continuous band of colors (Figure 3.8). U.S. amateur astronomer Henry Draper (1837–1882) first photographed a star's spectrum in 1872.

Stars are blazing balls of gas where many kinds of atoms emit light of all colors. Light from a star's bright visible surface is blurred into a continuous spectrum. As the light travels through the star's outer atmosphere, some of the colors (photons of certain wavelengths) are absorbed, producing dark absorption lines or dips in an intensity graph. These absorption lines identify the chemical elements that make up the star's atmosphere.

Refer to Figure 3.6. Identify the region of a star where (a) a continuous spectrum and (b) an absorption spectrum would originate. (a) _____ ; (b) _____

Answer: (a) Continuous; (b) absorption.

3.5 CHEMICAL COMPOSITION

Our Sun was the first star whose absorption spectrum was analyzed. In 1814, Bavarian physicist Joseph von Fraunhofer (1787–1826) recorded the strongest dark lines, now called **Fraunhofer lines**.

Figure 3.7. Astronomers identify iron in the Sun by matching dark lines in the Sun's absorption spectrum to a reference iron emission spectrum.

Since then astronomers have catalogued thousands of dark lines in the Sun's spectrum. By comparing these lines with the spectral lines produced by different chemical elements on Earth, they have found over 70 different chemical elements in the Sun (Figure 3.7).

How can the chemical composition of stars be determined? Assume that stars and their atmospheres are made of the same ingredients. _____

_____

Answer: By analyzing the dark lines in the star's spectrum and comparing them with those of each of the chemical elements on Earth.

3.6 SPECTRAL CLASSES

When you compare the spectra of stars like Polaris or Vega with the Sun's spectrum (Figure 3.8), you see that some look the same while others look quite different. Absorption spectra are used to classify stars into nine principal types, called **spectral classes**.

Hydrogen lines are much stronger in the spectra of some stars than in the Sun's spectrum. Astronomers once mistakenly thought that these stars had more hydrogen than other stars. They classified stars by the strength of the hydrogen lines in their spectra, in alphabetical order, from the strongest (called Class A) to the weakest (called Class Q).

U.S. astronomer Annie J. Cannon (1863–1941), who examined and classified the spectra of 225,300 stars, modified this classification system to its still-used form: O B A F G K M L T. (Astronomy students remember this order by saying: *"Oh Be A Fine Girl/Guy Kiss Me Love To."*)

All visible stars are roughly uniform in composition, made mostly of hydrogen and helium. U.S. astronomer Cecilia Payne-Gaposhkin (1900–1979) showed that the differences in the dark line patterns of stars are due primarily to their enormously different **surface temperatures**.

The sequence of spectral classes is a temperature sequence. The O stars are hottest, with the temperature continuously decreasing down to the coolest T stars. Each spectral class is arranged in 10 subclasses numbered 0 to 9, also in order of decreasing temperature.

Today discoveries necessitate extra classes. L and T were added recently for dwarfs cooler than class M stars.

What property determines the spectral class of a star? _____

Answer: Surface temperature.

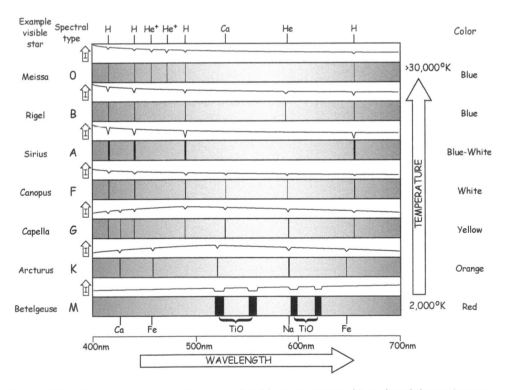

Figure 3.8. Seven main classes of spectra of visible stars, arranged in order of decreasing temperature. Intensity versus wavelengths of spectral lines shown in nanometers (nm). (He: neutral helium; H: hydrogen; Ca: calcium; Fe: iron; TiO: titanium oxide; Na: sodium; He+: helium ion.)

3.7 TEMPERATURE

The spectrum of a hot star and that of a cool star look very different. Examine Figure 3.8, which displays the spectral classes of visible stars. Each spectral class has key characteristics that serve, like numbers on a thermometer, to indicate a star's temperature.

Today's spectral classes of stars in order from highest to lowest temperatures, the approximate surface temperatures of these classes, and the main class characteristics are summarized in Table 3.1.

You can identify a new star's spectral class and probable temperature by comparing its spectrum to the images in Figure 3.8 and the class characteristics in Table 3.1.

TABLE 3.1   Spectral Class Characteristics

Spectral Class	Approximate Temperature (K)	Main Class Characteristics
O	>30,000	Relatively few lines; lines of ionized helium
B	10,000–30,000	Lines of neutral helium
A	7,500–10,000	Very strong hydrogen lines
F	6,000–7,500	Strong hydrogen lines; ionized calcium lines; many metal lines
G	5,000–6,000	Strong ionized calcium lines; many strong lines of ionized and neutral iron and other metals
K	3,500–5,000	Strong lines of neutral metals
M	2,000–3,500	Bands of titanium oxide molecules
L	1,300–2,000	Bands of iron hydride molecules
T	~700–1,300	Methane and water vapor

Figure 3.9. Star spectrums may be displayed as a graph of intensity versus wavelength or as a picture that looks like an old-fashioned photographic spectrum.

List the spectral class and probable temperature of each of the stars whose spectrum is shown in Figure 3.9. (a) _____ ; (b) _____

Answer: (a) A type (7500 K–10,000 K); (b) M type (<3500 K).

3.8 ORIGIN OF SPECTRAL CLASS CHARACTERISTICS

Atomic theory explains why hot blue (O-type) stars and cool red (M-type) stars produce spectra that look so different even though all stars are made of practically the same ingredients.

Every chemical element has a characteristic temperature and density at which it is most effective in producing visible absorption lines.

At extremely high temperatures, as in O stars, gas atoms are **ionized**, or broken up. Only the most tightly bound atoms such as singly ionized helium survive, and the lines of ionized atoms dominate the spectrum. When the temperature is around 5800 K, as in G stars such as our Sun, metal atoms such as iron and nickel remain neutral without being disrupted. At temperatures below 3500 K, as in M stars, even molecules such as titanium oxide can exist.

Does the absence of the characteristic absorption lines of a particular element like hydrogen in a star's spectrum necessarily mean that the star does not contain that element? Explain. _____

_____

Answer: No. The star's temperature determines which kinds of atoms can produce visible absorption lines.

3.9 MOTIONS

Stars have a **space velocity**, or motion through space with respect to the Sun, of many kilometers a second.

Space velocity has two components, which are measured independently: **radial velocity**, or speed toward or away from us along the line of sight; and **proper motion**, or the amount of angular change in a star's position per year (Figure 3.10).

A star's radial velocity is determined from analysis of its spectrum. The **Doppler shift** is an effect, discovered by Austrian physicist Christian Doppler (1803–1853), that applies to all wave motion. When a source of waves and an

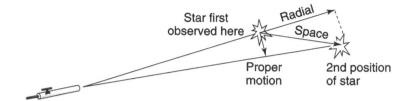

Figure 3.10. Space velocity has two components—radial velocity and proper motion.

observer are approaching or receding from each other, the observed wavelengths are changed.

A star's spectral lines (wavelengths) for any given element, such as iron, are compared with a reference spectrum. The star's wavelengths are shorter (**blueshift**) or longer (**redshift**) according to whether the star is moving toward or away from us (Figure 3.11).

The change in wavelength ($\Delta\lambda$) divided by the wavelength from a stationary source (λ) is proportional to the relative velocity (v) (unless v is comparable to the velocity of light, c). The formula is:

$$\frac{\Delta\lambda}{\lambda} = \frac{v}{c}$$

Proper motion is measured over an interval of 20 to 30 years. The average proper motion for all visible stars is less than 0.1 second of arc (0".1) per year. At that rate you won't notice any change in the appearance of your favorite constellation during your lifetime. But if you could return to observe the sky 50,000 years from now, it would look very different (Figure 3.12).

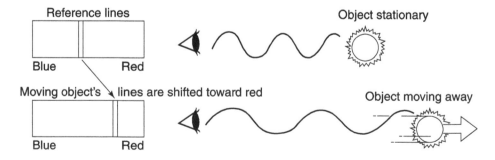

Figure 3.11. Doppler shift. A sky object's spectral lines for any given element are compared with reference lines. Redshifted spectral lines indicate that the object is moving away from us.

Today   50,000 years from now

Figure 3.12. Measured proper motion of the Big Dipper today indicates that the grouping will have a whole new look far in the future.

What is the angular change in an average visible star's position after 50,000 years? _____

Answer: 5000 seconds of arc, or 1.39° (almost three times the Moon's angular diameter, which is ½°).

Solution: 0".1 per year x 50,000 years = 5000"

3.10 OTHER PROPERTIES

Other information about stars is obtained from careful measurements of **spectral line shape**.

Gas **density**, the mass per unit volume, is indicated by **collisional broadening**. A broadened spectral line is produced when atoms collide more frequently in higher-density stars.

Axial rotation, the rotation of a star around its axis, is indicated by **rotational broadening**. If observable, a broadened line can yield a lower limit to the star's rate of rotation on its axis.

A splitting or broadening of spectral lines occurs in the presence of a **magnetic field**, a region where magnetic forces are detected, which is called the **Zm   n £ fe t**. The amount of splitting depends on the magnetic field strength.

These different kinds of broadening are not distinguishable to the unaided eye but are determined by careful analysis of the shape of the line using sensitive spectrometers.

List three properties of a star connected to its spectral line shape.
(1) _____ ; (2) _____ ; (3) _____

Answer: (1) Density; (2) axial rotation; (3) magnetic field strength.

3.11 DECODING A STAR'S SPECTRUM

Write a brief paragraph summarizing your understanding of how astronomers deduce different properties of a star from its spectrum. _____

_____

_____

_____

_____

Answer: Your summary should include the following concepts: (1) chemical composition, from the presence of the characteristic lines of certain elements; (2) temperature, from spectral type; (3) star's speed toward or away from us, from Doppler shift of the lines; (4) density, axial rotation, and surface magnetic fields, from line shape.

3.12 LUMINOSITY

Astronomers distinguish a star's **apparent brightness**—the way the star appears in the sky—from its **luminosity**—the actual amount of light a star shines into space each second.

The star we know best is our Sun. The luminosity of other stars is often stated in terms of the **Sun's luminosity** ($L_\odot$), which is 3.85×10^{26} watts. The Sun's luminosity is equivalent to 3850 billion trillion 100-watt light bulbs shining all together.

The most luminous stars are over a million times as luminous as the Sun. The dimmest stars known are less than 0.0001 as luminous as the Sun.

Deneb in Cygnus is about 60,000 times more luminous than the Sun.

Explain why the Sun looks much brighter to us than Deneb does.

_____

_____

_____

Answer: Deneb is 95 million times farther away from Earth (1500 light-years) than the Sun is (about 150 million km, or 93 million miles). A star's apparent brightness depends on both its luminosity and its distance from us.

3.13 PROPAGATION OF LIGHT

You cannot tell by looking at stars in the sky which ones have the greatest luminosity. The farther away (d) a star of luminosity (L) is, the less bright (B) it appears.

Light spreads out uniformly in all directions from a source so that the amount of starlight shining on a unit area falls off as 1 over the square of the distance away from the star. This relationship is called the **inverse square law** (Figure 3.13). The equation is: $B = L/4\pi d^2$.

Our Sun is exceptionally bright because it is so close to us. If it were located 100,000 times deeper in space, how many times fainter would it look?

Answer: 10 billion times fainter, or about like the brilliant blue-white star, Sirius.

Solution:

$$\frac{1}{(100,000)^2} = \frac{1}{10,000,000,000} \text{ as bright (or 10 billion times fainter)}$$

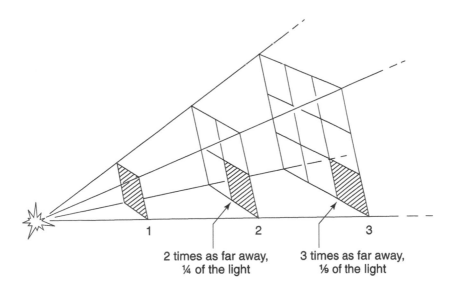

Figure 3.13.   Inverse square law. The same amount of starlight that shines on a square at 1 spreads out to illuminate four equal squares at 2 and nine equal squares at 3. Thus, if two stars have exactly the same luminosity but one is twice as far away from you as the other, the distant one will look only $\frac{1}{2}^2 = \frac{1}{4}$ as bright as the closer one, because you get one fourth the light in your eyes.

3.14 APPARENT MAGNITUDE ⭐

Apparent magnitude is a measure of how bright a star appears (see Section 1.7). The modern **magnitude scale** defines a first-magnitude star to be exactly 100 times brighter than a sixth-magnitude star.

This ratio agrees with the way our eyes respond to increases in the brightness of stars. What we see as a linear increase in brightness (a difference of one magnitude) is precisely measured as a geometrical increase in brightness (the fifth root of 100 or 2.512 times brighter).

Magnitude differences between stars measure the relative brightness of the stars. Table 3.2 lists approximate brightness ratios corresponding to sample magnitude differences.

Remember that the most negative magnitude numbers identify the brightest objects, while the largest positive magnitude numbers identify the faintest objects.

Refer to Tables 3.2 and 3.3. How much brighter does the Sun appear than Sirius? Explain.

_____

_____

Answer: 10 billion times brighter.

Solution: Magnitude difference is $(-26.7) - (-1.4) \cong 25$, corresponding to a brightness ratio of 10,000,000,000:1.

TABLE 3.2   Magnitude Differences and Brightness Ratios

Difference in Magnitude	Brightness Ratio
0.0	1:1
1.0	2.5:1
2.0	6.3:1
3.0	16:1
4.0	40:1
5.0	100:1
6.0	251:1
10.0	10,000:1
15.0	1,000,000:1
20.0	100,000,000:1
25.0	10,000,000,000:1

TABLE 3.3   Sample Magnitude Data

Subject	Description	Apparent Magnitude	Absolute Magnitude
Sun		–26.7	4.8
100-watt bulb	At 3 m (10 ft)	–18.7	66.3
Moon	Full	–12.5	32
Venus	At brightest	–4.7	28
Sirius	Brightest star	–1.4	1.5
Alpha Centauri	Closest seeable star	0	4.4
Andromeda Galaxy	Farthest seeable object	3.5	–21

3.15 ABSOLUTE MAGNITUDE

Absolute magnitude is a measure of luminosity, or how much light a star is actually radiating into space. If you could line up all stars at the same distance from Earth, you could see how they differ in intrinsic, or "true," brightness.

Astronomers define a star's absolute magnitude as the apparent magnitude the star would have if it were located at a standard distance of 10 parsecs from us. With the effects of distance canceled out, they can use absolute magnitude comparisons to determine differences in the actual light output of stars (Figure 3.14).

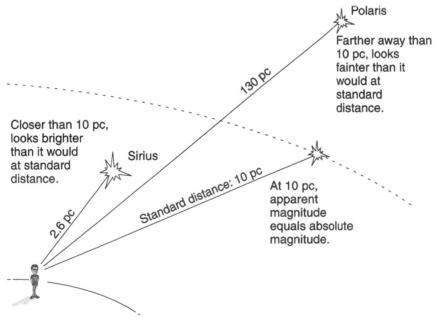

Figure 3.14.   Absolute magnitude and apparent magnitude.

If a star is farther than 10 parsecs from us, its apparent magnitude is numerically bigger than its absolute magnitude. (Large positive magnitude numbers indicate faint objects.) For example, Polaris is 130 pc away. Its apparent magnitude is +2.0, whereas its absolute magnitude is –4.1.

On the other hand, if a star is closer than 10 parsecs, its apparent magnitude is numerically smaller than its absolute magnitude. Thus, Sirius is 2.6 pc away. Its apparent magnitude is –1.4, whereas its absolute magnitude is only +1.5.

Consider the two bright stars Deneb and Vega. Refer back to Table 1.1 to fill in the chart below. Then tell (a) which looks brighter? _____ (b) Which is really more luminous? _____ (c) What factor makes your answers to (a) and (b) different? _____

Two Bright Stars

Star	Constellation	Apparent Magnitude	Absolute Magnitude
Deneb	Cygnus	(a)	(b)
Vega	Lyra	(c)	(d)

Answer: Chart: (a) 1.25; (b) –7.5; (c) 0.03; (d) 0.6.

(a) Vega (numerically smaller apparent magnitude). (b) Deneb (numerically more negative absolute magnitude). (c) The distance the stars are from us.

3.16 DISTANCES FROM MAGNITUDES

The difference between the apparent magnitude (m) and absolute magnitude (M) is called the **distance modulus (m – M)**. In formula form:

$$m - M = 5 \log \left(\frac{\text{distance in parsecs}}{10} \right)$$

A star's apparent magnitude can be measured directly. For a distant star whose parallax cannot be measured but whose absolute magnitude is known, as from consideration of its spectrum, the distance modulus can be used to calculate distance.

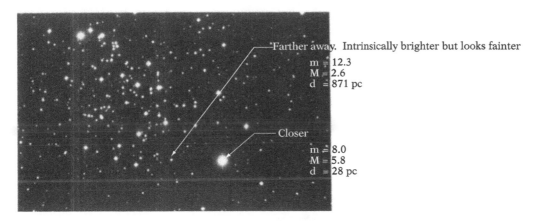

Figure 3.15. Star clusters, showing difference in apparent and absolute magnitude. Farther away (intrinsically brighter but looks fainter): m = 12.3; M = 2.6; d = 871 pc. Closer: m = 8.0; M = 5.8; d = 28 pc.

Refer to Figure 3.15. Give the distance modulus of the stars that are (a) closer _____ ; (b) farther away _____.

Answer: (a) 2.2; (b) 9.7.

3.17 COMPARISONS

Be sure you understand the ideas presented so far by answering the following questions about four of the Sun's neighbor stars described in the chart below.

Four Nearby Stars

Star	Apparent Magnitude	Absolute Magnitude	Spectral Class	Parallax (in ")
Alpha Centauri A	–0.0	4.3	G	0.742
Thuban	4.7	5.9	K	0.173
Barnard's Star	9.5	13.2	M	0.549
Altair	0.8	2.1	A	0.194

Which star is (a) hottest? _____ (b) coolest? _____
(c) brightest looking? _____ (d) faintest appearing? _____
(e) intrinsically (actually) most luminous? _____ (f) intrinsically least

luminous? _____ (g) closest? _____ (h) most distant? _____ Explain your answers.

Answer: (a) Altair, spectral class A; (b) Barnard's Star, spectral class M; (c) Alpha Centauri A, apparent magnitude –0.0; (d) Barnard's Star, apparent magnitude 9.5; (e) Altair, absolute magnitude 2.1; (f) Barnard's Star, absolute magnitude 13.2; (g) Alpha Centauri A, parallax = 0".742, or distance = 1/parallax = 1/0".742 = 1.3 pc; (h) Thuban, parallax = 0".173, or distance = 1/0".173 = 5.7 pc.

3.18 HERTZSPRUNG–RUSSELL DIAGRAM

A basic link between luminosities and temperatures of stars was discovered early in the twentieth century by two independent astronomers, Henry N. Russell (1877–1957) of the U.S. and Ejnar Hertzsprung (1893–1967) of Denmark. The **Hertzsprung–Russell (H–R) diagram** is a plot of luminosity versus temperature. Astronomers use the H–R diagram widely to check their theories (Figure 3.16).

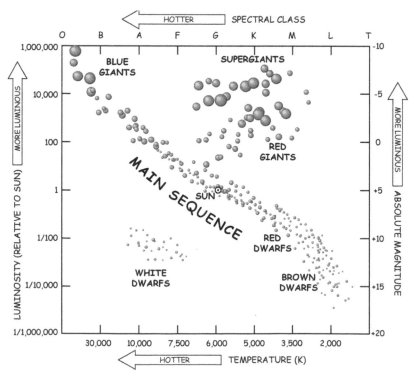

Figure 3.16. Hertzsprung-Russell (H–R) diagram for many stars. Temperature increases from right to left. Luminosity increases from bottom to top.

Every dot on an H–R diagram represents a star whose temperature (spectral class) is read on the horizontal axis and whose luminosity (absolute magnitude) is read on the vertical axis.

Significantly, when a few thousand stars are chosen randomly and plotted on an H–R diagram, they fall into definite regions. This pattern indicates that a meaningful connection exists between a star's luminosity and its temperature. Otherwise, the dots would be scattered randomly all over the graph.

About 90 percent of the stars lie along a band called the **main sequence**, which runs from the upper left (hot, very luminous **blue giants**) across the diagram to the lower right (cool, faint **dwarfs**). Red dwarfs are the most common type of nearby star.

Most of the other 10 percent of stars fall into the upper right region (cool, bright **giants** and **supergiants**) or in the lower left corner (hot, low-luminosity **white dwarfs**).

Identify the location of the following stars indicated on the H–R diagram in Figure 3.17. Each star's absolute magnitude is given in parentheses. Refer to Table 1.1 and Appendix 5 spectral class. (a) Rigel (–6.6) _____ ;

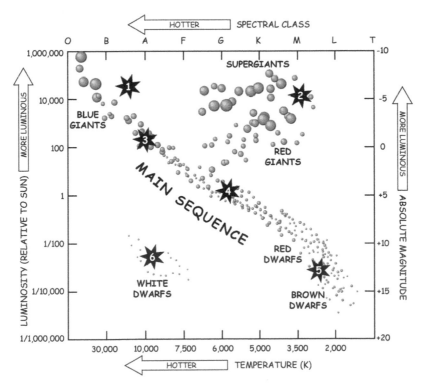

Figure 3.17. An incomplete H–R diagram for selected stars.

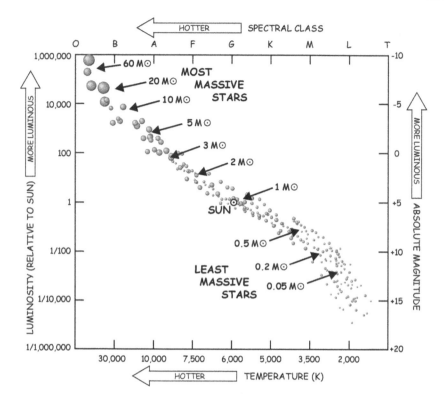

Figure 3.18. Masses of some typical main sequence stars. (M$_\odot$ = mass of Sun.)

(b) Vega (0.6) _____ ; (c) Sun (4.8) _____ ; (d) Betelgeuse (−5.0) _____ ; (e) Barnard's Star (13.2) _____ ; (f) Sirius B (11.3) _____

Answer: (a) 1; (b) 3; (c) 4; (d) 2; (e) 5; (f) 6.

3.19 MASS–LUMINOSITY RELATION

A star's position on the main sequence is determined by its **mass**, or the amount of matter the star contains.

The main sequence is a sequence of stars of decreasing mass, from the most massive, most luminous stars at the upper end to the least massive, least luminous stars at the lower end (Figure 3.18).

An empirical **mass–luminosity relation** for main sequence stars, found from binary stars, says that the more massive a star is, the more luminous it is. The luminosity of a star is approximately proportional to its mass raised to the 3.5 power.

The **mass of the Sun, M₀** = 2 x 10³⁰ kg, is practically 333,000 times the mass of Earth. Stellar masses do not vary enormously along the main sequence as stellar luminosities do. The faintest red dwarfs have a mass about one tenth of the Sun's. (A gas object with a mass between 1/100 and 1/10 the Sun's, called a **brown dwarf**, may shine briefly but is too small to get hot enough to become a star.) The largest mass of a stable star is about 60 to 75 times that of the Sun.

What basic property of a star determines its position on the main sequence of the H–R diagram; that is, what determines its luminosity and temperature?

_____

Answer: Mass.

3.20 SIZES AND DENSITIES ✦

Our Sun is the only star that is close enough to allow astronomers to measure its size directly.

The diameter of the Sun is 1.39 million km (about 864,000 miles). That is equal to 109 Earths placed next to each other.

If the absolute temperature and luminosity of a star are determined, then the size of the star can be calculated from the **Stefan–Boltzman radiation law**. This law says that the luminosity (L) of a star is proportional to the square of its radius (R) times the fourth power of its surface temperature (T). The equation is:

$$L = 4\pi R^2 \sigma T^4$$

where σ is the Stefan–Boltzmann constant (Appendix 2).

Stars on the main sequence vary in size continuously from the large blue-white giants, about 25 times the Sun's *radius* (R_0), down to the common cool red dwarfs, which are only about 1/10 the Sun's radius.

The largest stars are the supergiants such as Betelgeuse in Orion, whose radius is about 400 times the Sun's radius. You could fit more than a million stars like our Sun inside Betelgeuse! The smallest common stars are the white dwarfs, roughly the size of Earth.

The mean **density**, or the mass per unit volume, of the Sun is 1.4 g/cm³, slightly more than that of water. Red giant stars and white dwarf stars both have about the same mass as the Sun but are very different sizes.

What can you say about the densities of red giants and white dwarfs compared to the Sun? Explain. _____

_____

_____

_____

_____

_____

Answer: Red giants have very low density compared to the Sun. They have the same amount of mass in a much bigger volume. (Their mean density is about the same as that of a vacuum here on Earth.)

White dwarfs are extremely dense. They have the same amount of mass packed into a much smaller volume. (One teaspoon of material from a white dwarf would weigh several tons on Earth.)

3.21 DOUBLE STAR SYSTEMS

Many stars that look single to the unaided eye are not. A **binary star** is formed by a pair of stars that revolve around a common center of gravity as they travel through space together (Figure 3.19). The masses of the stars can be figured from the angular size and period of their orbits.

Binary stars are classified by the way they are observed.

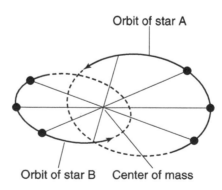

Orbit of star A

Orbit of star B   Center of mass

Figure 3.19. A binary star consists of two stars A and B orbiting a common center of mass, bound together by their mutual gravity.

A **visual binary** can be resolved with a telescope so that two separate stars can be seen. Over 70,000 visual binaries are known. Mizar in Ursa Major was the first binary star discovered, in 1650. Beautiful Albireo in Cygnus is a colorful yellow and blue star. You can see these and many others in a small telescope. (See Useful Resources for observers' guides.)

Many visible stars may have companions that are too faint to be seen. An **astrometric binary** is a visible star plus an unseen companion star. The presence of an unseen companion is inferred from variable proper motion of the visible star. Brilliant Sirius (Sirius A) in Canis Major was an astrometric binary from 1844, when its nature was detected, until 1862. Then its faint companion star (Sirius B) was observed.

A **spectroscopic binary** cannot be resolved in a telescope. Its binary nature is revealed by its spectrum. A varying Doppler shift is apparent in the spectral lines as the stars approach and recede from Earth. Almost a thousand spectroscopic binaries have been analyzed. The brighter member of Mizar (Mizar A) is a spectroscopic binary.

An **eclipsing binary** is situated so that one star passes in front of its companion, cutting off light from our view at regular intervals. An eclipsing binary regularly changes in brightness. You can see the famous eclipsing binary Algol, the Demon, in Perseus. Algol "winks" from brightest magnitude 2.2 to least bright magnitude 3.5 in about 2 days and 21 hours.

An **optical double** is a pair of stars that appear to be close to each other in the sky when viewed from Earth. Actually, one is much more distant than the other, and they have no physical relationship to one another.

Test your eyesight by finding both Mizar and Alcor (nicknamed "the testers"), the optical double in the handle of the Big Dipper in Ursa Major.

How does an optical double differ from a visual binary? _____

_____

_____

Answer: The stars in an optical double are far apart and have no actual relationship to one another. The stars in a visual binary are bound together in space by their mutual gravity.

SELF-TEST

This self-test is designed to show you whether or not you have mastered the material in Chapter 3. Answer each question to the best of your ability. Correct answers and review instructions are given at the end of the test.

1. Refer to the chart on page 83. From the measured parallax, find the distance to Barnard's Star in (a) parsecs _____ ; (b) light-years _____ .

2. Explain why the bright (dark) spectral lines of light emitted from (absorbed by) the atoms of an element are unique to that element. _____

   _____

   _____

   _____

3. Explain how a star's spectrum is formed. _____

   _____

   _____

   _____

4. List the following types of spectral lines in order as they appear in stars of decreasing temperature.

   ____ (1) Very strong hydrogen lines.   ____ (4) Neutral helium.
   ____ (2) Ionized helium.   ____ (5) Neutral metals.
   ____ (3) Bands of titanium oxide   ____ (6) Ionized metals.
            molecules.

5. Match the following properties of a star that can be deduced from its spectrum with the appropriate method listed on the right.

   ____ (a) Chemical composition.   (1) Doppler shift.
   ____ (b) Temperature.   (2) Spectral type (class).
   ____ (c) Radial velocity.   (3) Line shape.
   ____ (d) Gas density, axial rotation,   (4) Characteristic lines.
            magnetic field.

6. The proper motion of Sirius is 1.34" per year. Find how much Sirius will change its position on the celestial sphere in the next 1000 years._____

_____

7. Define space velocity. _____

_____

8. Refer to Table 1.1. By using their apparent magnitudes, absolute magnitudes, and spectral classes, match one of the four stars to each description.

____ (a) Hottest.   (1) Betelgeuse.
____ (b) Coolest.   (2) Procyon.
____ (c) Most luminous.   (3) Spica.
____ (d) Least luminous.   (4) Sirius.
____ (e) Brightest.
____ (f) Faintest.
____ (g) Closest.
____ (h) Most distant.

9. Label the following on the H–R diagram in Figure 3.20:
   (1) Surface temperature of star (K).   (7)   White dwarfs.
   (2) Absolute luminosity (Sun = 1).   (8)   Supergiants.
   (3) Spectral class.   (9)   Blue giants.
   (4) Absolute magnitude.   (10) Red dwarfs.
   (5) Main sequence.   (11) Brown dwarfs.
   (6) Red giants.

10. What is the most basic property of a star that determines its location on the main sequence (its temperature and luminosity)? _____

11. Use the H–R diagram to explain why, compared to our Sun, red giants must be very large and white dwarfs must be very small. _____

_____

_____

_____

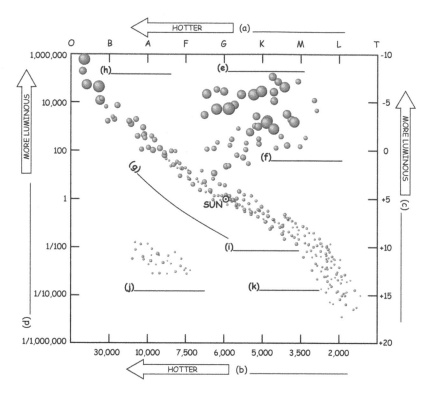

Figure 3.20. An incomplete H–R diagram.

____ (a) Can be resolved with a telescope.

____ (b) Unseen companion inferred from variable proper motion of visible companion.

____ (c) Binary nature revealed by its spectrum.

____ (d) Changes in brightness regularly as one star blocks its companion from our view.

____ (e) Member stars have no actual physical relationship to one another.

(1) Astrometric binary.
(2) Eclipsing binary.
(3) Optical double.
(4) Spectroscopic binary.
(5) Visual binary.

ANSWERS

Compare your answers to the questions on the self-test with the answers given below. If all of your answers are correct, you are ready to go on to the next chapter. If you missed any questions, review the sections indicated in parentheses following the answer. If you missed several questions, you should probably reread the entire chapter carefully.

1. (a) 1.8 pc;   (b) 6.0 ly.   (Section 3.1)

Solution: Measured parallax is

$$0".549 \text{ and } \frac{1}{0".549} = 1.8 \text{ pc}$$

2. Each spectral line is light of a particular wavelength emitted (or absorbed) by the atom when one of its electrons jumps between a higher and a lower energy level (orbit). Each element has its own unique set of allowed electron orbits, so each element has its own characteristic set of spectral lines.   (Sections 3.2, 3.3)

3. Stars are blazing balls of gas where many kinds of atoms emit light of all colors. This light, emitted from the star's surface, passes through the star's outer atmosphere. There, atoms of each element absorb their characteristic wavelengths, so a pattern of dark lines crosses the continuous band of colors—the star's spectrum.   (Sections 3.3, 3.4)

4. 2;   4;   1;   6;   5;   3.   (Section 3.7)

5. (a) 4;   (b) 2;   (c) 1;   (d) 3.   (Sections 3.3, 3.5 through 3.10)

6. 1340", or roughly one third of a degree.

Solution: Proper motion = 1.34" per year. A degree is equal to 3600".

(1.34" per year) x 1000 years   (Section 3.9)

7. Velocity of a star with respect to the Sun.   (Section 3.9)

8. (a) 3;   (b) 1;   (c) 1;   (d) 2;   (e) 4;   (f) 3;   (g) 4;   (h) 1.
(Sections 3.6, 3.7, 3.12 through 3.16)

Solution:

	Spectral Class	Distance (ly)	Apparent Magnitude	Absolute Magnitude
Betelgeuse	M	522	0.45	−5.0
Procyon	F	11	0.40	2.8
Spica	B	262	0.98	−3.6
Sirius	A	9	−1.44	1.5

9. (a) 3; (b) 1; (c) 4; (d) 2; (e) 8; (f) 6; (g) 5; (h) 9;
   (i) 10; (j) 7; (k) 11.   (Section 3.18)

10. Mass.   (Section 3.19)

11. Red giants are relatively cool but luminous; hence, they must have a large surface area radiating energy. White dwarfs are relatively hot but faint; hence, they must have a small surface area radiating energy into space.   (Sections 3.18 through 3.20)

12. (a) 5; (b) 1; (c) 4; (d) 2; (e) 3.   (Section 3.21)

4

THE SUN

Thou dawnest beautifully in the horizon of the sky
O living Aton who wast the Beginning of life!

Akhenaton (c. 1386–1358 B.C.)
"Hymn to the Sun"

Objectives

☆ List some reasons why modern astronomers study the Sun.

☆ Define the solar constant, and explain why it is important to know if it is truly constant with time.

☆ Define the astronomical unit, AU.

☆ Relate the formation, properties, and motions of the Sun as a star.

☆ Sketch the structure of the Sun and identify the corona, chromosphere, photosphere, convection zone, radiation zone, and core.

☆ Describe the Sun's rotation and magnetic field.

☆ List the basic physical dimensions of the Sun.

☆ Describe some modern tools and techniques for studying the Sun.

☆ Describe the origin, properties, and cyclic nature of sunspots, and explain how sunspot variations are related to solar activity.

☆ Compare and contrast the origin and nature of solar granules, spicules, faculae, flares, coronal mass ejections, and prominences.

☆ Describe the origin and nature of the solar wind.

☆ Identify and describe some ways that violent solar eruptions impact Earth.

☆ Outline the goal and results of the solar neutrino experiments and helioseismology.

4.1 SUN AND EARTH

The Sun is the star closest to Earth. It provides the light, heat, and energy for life.

Ancient peoples worshipped the Sun as a life-giving god. Some of the names given to the Sun god were Aton, Apollo, Helios, and Sol. Scientists study the Sun today. It is critical to Earth and is a key to understanding distant stars that cannot be observed in detail.

The Sun's total energy output is enormous. The **Sun's luminosity L_0** is 3.85×10^{26} watts. Solar energy is practically inexhaustible. The amount of the Sun's energy that falls per second on Earth's outer atmosphere, called the **solar constant**, is about 1400 watts/m^2 (126 watts/square foot). This amount of energy provides about as much heat and light in a week as is available from all of our known reserves of oil, coal, and natural gas.

Our Sun is dynamic and seething (Figure 4.1). It is in turn extraordinarily

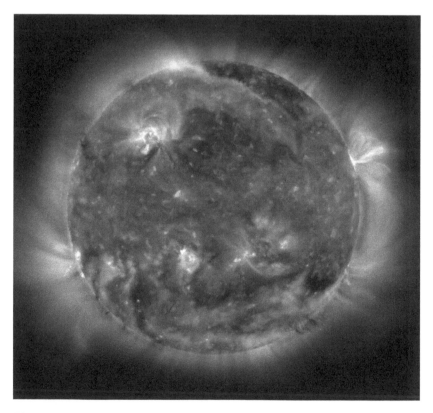

Figure 4.1. Solar activity. Hot active regions on the Sun imaged by an extreme ultraviolet imaging telescope aboard European/U.S. robot Solar and Heliospheric Observatory (SOHO).

THE SUN ★ 97

active and relatively quiet. Changes in solar energy output affect Earth's climate, atmosphere, and weather, as well as modern power-transmission and communications systems. These changes are monitored to learn exactly how the Sun affects Earth.

State three reasons why modern astronomers, physicists, and engineers are using their most sophisticated techniques to determine the true nature of the Sun.

(1) _____

_____

(2) _____

_____

(3) _____

_____

Answer: (1) The Sun is an almost inexhaustible source of present and potential future energy. It is also free and nonpolluting! (2) The Sun is the only star close enough to observe in detail, so astronomers use it to determine what other stars are like. (3) Changes in the Sun's energy output affect Earth's climate, atmosphere, and weather, as well as power-transmission and communications systems.

4.2 DISTANCES AND SIZE

The average distance between Earth and the Sun, called the **astronomical unit** (**AU**), is about 150 million km (93 million miles).

Astronomers calculate this distance from planetary data obtained by radar ranging. They use the astronomical unit as a measure of distance in the solar system (Table 8.1).

The Sun is a huge gaseous sphere. We see the apparent surface layer in the sky. Its **radius** ($R_{\odot}$) is about 696,000 km (432,000 miles). From Earth, the **Sun's angular diameter** = 32′ (about ½°) looks deceptively equal to the full Moon's. This illusion occurs because the Sun is 400 times farther away.

WARNING: *You could permanently blind your eyes if you observe the Sun without first taking proper precautions!* You should *never* look at the Sun directly and *never* look at the Sun through an optical instrument unless special solar filters properly cover the full aperture.

A way to observe the Sun is to project its image onto a screen and look only at the Sun's image on the screen (Figures 4.2 and 4.3).

About how many minutes does it take for sunlight to travel 1 AU? *Tip*: distance = speed x time. You can rewrite this: time = distance/speed. _____

Answer: About 8.3 minutes. (That means that if the Sun stopped shining, you would not know about it until 8.3 minutes later.)

Solution: The speed of light is ≈ 300,000 km (186,000 miles) per second and

$$1 \text{ AU} = 150{,}000{,}000 \text{ km}$$

$$\frac{150{,}000{,}000 \text{ km}}{300{,}000 \text{ km/second}} \qquad \frac{93{,}000{,}000 \text{ miles}}{186{,}000 \text{ miles/second}}$$

$$= 500 \text{ seconds or } 8.3 \text{ minutes}$$

Figure 4.2. Projected image of the Sun focused on a screen behind the eyepiece of a small reflecting telescope. A way to avoid looking at the Sun while pointing a telescope is to use the shadow of the telescope cast on the screen as a guide.

Figure 4.3. Projected image of the Sun focused on a screen behind the eyepiece of a small refracting telescope. A way to avoid looking at the Sun while pointing a telescope is to use the shadow of the telescope cast on the screen as a guide.

4.3 MAKEUP

The **nebular theory**, first proposed by German philosopher Immanuel Kant (1724–1804), says that our Sun and its planets formed together from a rotating cloud of interstellar gas and dust called the **solar nebula** about 4.6 billion years ago.

The solar nebula condensed into the newly forming Sun encircled by a rotating disk of gas and dust out of which the planets, moons, and other solar system objects formed (Figure 4.4). The Sun's mass, 2 x 1030 kg—more than 99 percent of the solar system's—provides the gravitational force that keeps the planets circling it. Its surface gravity is 28 times Earth's.

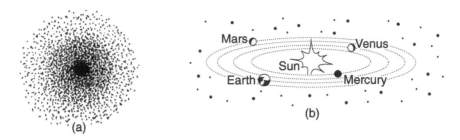

Figure 4.4. Solar nebula theory. (a) A rotating nebula condensed to our Sun surrounded by a contracting disk where (b) the planetary system was born.

More than 70 **chemical elements** have been identified in the Sun's spectrum. The Sun's outer layers likely have the same chemical composition as the Sun had at birth: about 71 percent hydrogen, 27 percent helium, and 2 percent other elements by mass. The Sun's core probably has subsequently changed to over 50 percent helium in nuclear fusion reactions.

Why do astronomers expect to find many other stars that have planets circling around them? _____

_____

_____

Answer: The nebular theory says that the planets circling the Sun were born together with their star. Since the Sun is a typical star, it seems likely that many other stars were also born together with a family of planets.

4.4 THE SUN'S STRUCTURE

Our picture of the structure of the Sun (and other stars) comes from direct observations of its outer layers plus indirect theoretical calculations of the behavior of gases deep inside that we cannot see.

The three outer layers are called the Sun's **atmosphere**.

The **photosphere**, from the Greek "light ball," is the visible surface of the Sun. The photosphere is a hot, 500-km (300-mile), opaque gas layer about 5800 K (10,000° F) from which energy is radiated into space. The **limb** is the apparent edge of the Sun's disk. It looks darker than the center, an effect called **limb darkening**, because light from the limb comes from higher, cooler regions of the photosphere.

The **chromosphere**, from the Greek "color ball," is a thin, transparent layer that extends about 10,000 km (6000 miles) above the photosphere. It is normally visible from Earth only during a total eclipse of the Sun, when it glows red due to its hydrogen gas. The temperature unexpectedly increases outward through the chromosphere, where the average temperature of matter is about 15,000 K.

The **corona**, from the Latin "crown," is the outermost atmosphere just above the chromosphere. It is a rarified, hot gas that extends many millions of kilometers into space. Because of its high temperature—up to 2 million K in the outer part—the corona shines bright at X-ray wavelengths. During a total eclipse of the Sun, it is strikingly visible as a jagged white halo around the briefly hidden photosphere (Figure 4.5).

Below the photosphere is the Sun's **interior**. Theorists figure that temperature and density increase inward from the surface. No known element can survive as a solid or liquid at the extremely high solar temperatures. So the Sun must be made of very hot ionized gases (plasma) throughout.

Deep inside, the temperature must rise to 15 million K, the pressure to 200 billion atmospheres, and the density to over 100 times that of water. The

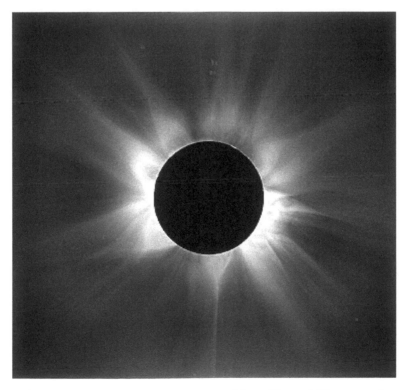

Figure 4.5. The corona reaches outward for millions of kilometers during a total eclipse of the Sun.

core is the power plant where nuclear fusion reactions generate the Sun's energy (Section 5.5). There, hydrogen is fused into helium.

The intense energy released in the core provides heat inside the Sun and enough pressure to balance the inward pull of gravity. It is slowly transmitted outward. Photons are repeatedly absorbed and re-emitted at lower energies in the crowded **radiation zone**.

From there, circulating currents of gas in the **convection zone** transfer most of the energy as heat to the outer layers. It takes about 20 million years for energy produced in the core to surface and become sunshine.

Identify the regions of the Sun lettered on Figure 4.6. (a) _____ ; (b) _____ ; (c) _____ ; (d) _____ ; (e) _____ ; (f) _____

Answer: (a) Corona; (b) chromosphere; (c) photosphere; (d) convection zone; (e) radiation zone; (f) core.

Figure 4.6. Regions of the Sun.

4.5 ROTATION

The Sun keeps turning around its axis in space, from west to east, as Earth does. But there is a difference. All of Earth makes a complete turn in a day. The whole Sun does not turn around together at the same rate.

The **period of rotation**, or the length of time for one complete turn, is fastest at the Sun's equator (about 25 days), slower at middle latitudes, and slowest at the poles (about 35 days). This strange rotation pattern probably contributes to the violent activity that takes place on the Sun, described in the sections that follow.

How is it possible for different parts of the Sun to rotate at different rates, in contrast to Earth, all of which makes a complete turn in a day? _____

_____

Answer: The Sun is a gaseous sphere and not a rigid solid as is Earth.

4.6 DATA

Summarize the data you have on the Sun's properties by filling in the convenient reference Table 4.1.

TABLE 4.1 Properties of the Sun

Quantity	Method of Measurement	Value
(a) Average distance from Earth	Radar ranging of planets	_____
(b) Angular diameter in sky	Solar telescope	_____
(c) Diameter	Angular diameter and distance	_____
(d) Mass	Planets' orbital motions	_____
(e) Average density	Mass and volume	_____
(f) Solar constant (solar energy incident on Earth)	High-altitude aircraft	_____
(g) Luminosity	Solar constant and distance from Earth	_____
(h) Surface temperature	Luminosity and radius	_____
(i) Spectral type	Spectrograph	_____
(j) Apparent magnitude	Photometer	_____
(k) Absolute magnitude	Apparent magnitude and distance from Earth	_____
(l) Rotation period	Sunspots' motions; Doppler shift	_____
(m) Chemical composition of outer layers	Sun's absorption spectrum	_____
(n) Surface gravity	Mass and radius	_____

Answer: (a) About 150 million km (93 million miles); (b) 32'; (c) 1,390,000 km (864,000 miles); (d) 2 x 10^{30} kg; (e) 1.4 g/cm^3; (f) 1400 watts/m^2 (126 watts/ft^2); (g) 3.85 x 10^{26} watts; (h) about 5800 K; (i) G2; (j) –26.75; (k) 4.8; (l) equator: about 25 days; poles: about 35 days; (m) outer layers: about 71 percent hydrogen, 27 percent helium, 2 percent more than 70 other elements by mass; (n) 28 times Earth's or 274 m/s2.

4.7 OBSERVATIONS

Astronomers are using sophisticated tools and techniques to observe the Sun more closely and in more detail than ever before.

Optical solar telescopes at the U.S. National Solar Observatory ▶www. nso.edu◀ and worldwide continually image the Sun's visible surface with its

Figure 4.7a. Building the 2-ton, 25-foot wide (with solar panels) Solar and Heliospheric Observatory (SOHO). Lower Service Module gives power, thermal control, pointing, and telecommunications. Upper Payload Module holds scientific instruments.

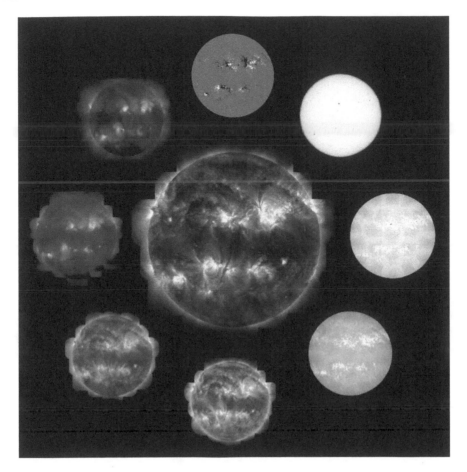

Figure 4.7b. Sun in different wavelengths. Central image, ultraviolet. Clockwise from top: magnetic map, white light, five ultraviolet, and X-ray.

changing features. Arrays of giant **radio telescopes** receive and record radio waves. **Infrared telescopes** observe the solar limb and map sunspots.

In space, instruments monitor the Sun in all parts of the electromagnetic spectrum to detect solar features, radiations, particles, and fields normally blocked by Earth's atmosphere (Figure 4.7a). **Ultraviolet, X-ray, and gamma ray telescopes** on spacecraft record images of processes in the hottest and most active regions of the Sun (Figure 4.7b).

Spectroheliographs image the Sun in light of essentially a single wavelength belonging to one gas such as hydrogen or calcium. The monochromatic images are called **spectroheliograms**. Because hot gases produce light at specific wavelengths, the data reveal local surface temperatures and phenomena (Figure 4.8).

Formerly, the Sun's chromosphere and corona could be observed directly only during the few minutes of a total eclipse of the Sun when the much brighter photosphere was hidden. Now astronomers do not have to wait for one of these rare natural events to occur. They use a **coronagraph**, a telescope equipped with a disk that blocks light from the photosphere, to create an artificial eclipse at will.

The corona is very hot and dynamic, frequently discharging high-energy radiation and mass that impact life on Earth. Several spacecraft are designed to help us understand and predict the Sun's activity. They stream back spectacular images and data.

ESA/NASA robot Solar and Heliospheric Observatory (SOHO) (1995–) continuously records the Sun's atmosphere, wind, and interior. SOHO's

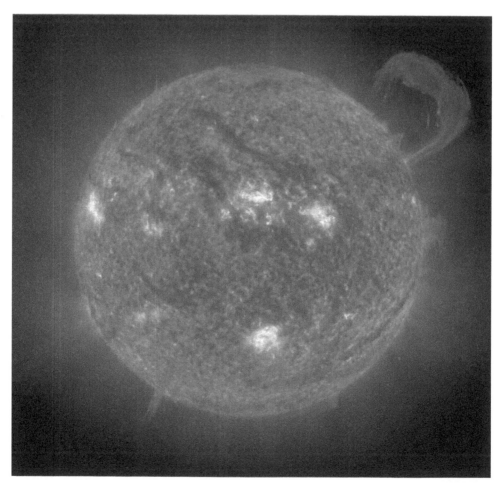

Figure 4.8. The Sun with a big loop prominence, imaged in light of a spectral line at 30.4 nm, which shows the upper chromosphere at about 60,000 K.

view is unimpeded in orbit around the Sun at Lagrangian point L1, 1.5 million km (1 million miles) sunward from Earth, where the gravitational forces of both are equal. ▶sohowww.nascom.nasa.gov◀

U.S. Solar Dynamics Observatory (SDO) (2010–) measures solar oscillations, magnetic fields, corona hot plasma, wind, and irradiance, or total light energy output measured at Earth, in several wavelengths simultaneously. Unlike many spacecraft, SDO does not have a recording system to store data on until it passes over a ground receiving station. Instead, SDO operates in a geosynchronous orbit, a geocentric orbit with an orbital period equal to Earth's sidereal rotation period. It is always above and can download to a specially built station in New Mexico, U.S. ▶http://sdo.gsfc.nasa.gov◀

While in orbit around the Sun at L1, U.S. robot Genesis (2001–2004) collected some 10 to 20 micrograms (comparable to a few grains of salt) of solar wind particles and returned its prize to Earth. Now scientists can measure precisely the composition of actual material from the Sun. ▶http://genesis mission.jpl.nasa.gov◀

First to orbit nearly perpendicular to the ecliptic plane, European/U.S. robot Ulysses (1991–2009) imaged the Sun's magnetic fields, radiation and particles, and environment at all solar latitudes. Ulysses flew over the polar regions at minimum solar activity in 1994–1995, at maximum solar activity in 2000–2001, and a third time at the start of the latest solar cycle. ▶http://ulysses .jpl.nasa.gov◀

Why do different features of the Sun appear in pictures taken in light of different wavelengths such as visible light, ultraviolet rays, or X-rays? *Tip*: Review Section 2.10 if necessary. _____

Answer: Different wavelengths are produced in regions of different temperatures where different conditions and activities prevail.

4.8 SEETHING SURFACE

Optical telescopes reveal that the photosphere has a grainy appearance, called **granulation**. Bright spots that look like rice grains, called **granules**, dot the Sun's disk in high-resolution images (Figure 4.9).

Granules, cells up to 1000 km (625 miles) across, are the tops of rising currents of hot gases from the convection zone. Individual granules last an average of 5 minutes each. They look brighter than neighboring dark areas

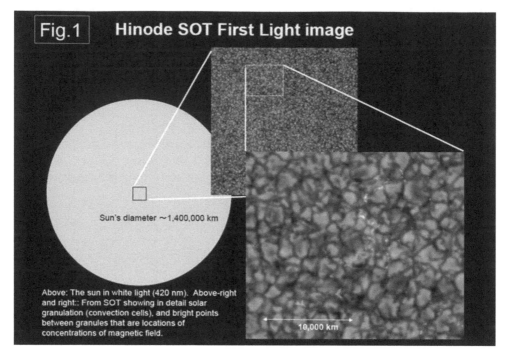

Above: The sun in white light (420 nm). Above-right and right:: From SOT showing in detail solar granulation (convection cells), and bright points between granules that are locations of concentrations of magnetic field.

Figure 4.9. Solar granules. Bright points between granules represent locations of magnetic field concentrations.

because they are about 300° hotter. The dark areas are descending currents of cooler gases.

Granules belong to **supergranules**, which are large, organized convection cells up to 30,000 km (19,000 miles) across, on the Sun's disk. Supergranules last several hours. They have a flow of gases from their centers to their edges, in addition to the vertical gas currents in the granules.

Spicules, jets of gas up to 10,000 km (6000 miles) tall and 1000 km (600 miles) across, rise like fiery spikes into the chromosphere around the edges of supergranules. They change rapidly and last about 5 to 15 minutes.

Bright, white surface patches, called **faculae**, from the Latin "little torches," may be visible near the Sun's limb. Their appearance seems to signal coming solar activity.

What causes granulation? _____

Answer: Gases rising from the Sun's hot interior.

4.9 SUNSPOTS

Sunspots are temporary, dark, relatively cool blotches on the Sun's bright photosphere. They usually appear in groups of two or more. Individual sunspots last anywhere from a few hours to a few months.

The largest sunspots are visible at sunrise or sunset or through a haze. Observations of sunspots were first recorded in China before 800 B.C.

A typical sunspot is roughly twice as big as Earth. The largest sunspots may be bigger than 10 Earths.

Sunspots really shine brighter than many cooler stars. They look dark only in comparison to the hotter, dazzling surrounding photosphere. The temperature is about 4200 K in the **umbra**, or core. The **penumbra**, or outer gray part of a large spot, is a few hundred degrees cooler than the photosphere.

Frequently sunspots appear in groups, or **solar active** regions, where the most violent solar activity occurs. The first telescopic observations of sunspots and their motions, reported by Galileo in 1610, had an historic impact (Section 8.7). Galileo correctly concluded that the Sun's rotation carries sunspots around.

Figure 4.10. Umbra and penumbra of a sunspot, with granulation in the surrounding photosphere.

Identify the umbra, penumbra, and photosphere, lettered on Figure 4.10, and indicate the approximate temperature of the umbra. (a) _____ ; (b) _____ ; (c) _____

Answer: (a) Photosphere; (b) penumbra; (c) umbra, 4200 K.

4.10 ACTIVITY CYCLES

At any one time more than 300 sunspots—or none at all—may appear on the Sun's disk. The number of sunspots regularly rises to a maximum and falls to a minimum in an approximately 11-year cycle, called the **sunspot cycle**.

The sunspot cycle is watched carefully from Earth because it marks the **solar activity cycle**. The Sun is most active with greatest outbursts of energy and radiation for about 4.8 years, during which sunspots are increasingly numerous. After **sunspot maximum**, the number of sunspots decreases for about 6.2 years to a **sunspot minimum** as solar activity lessens. The current cycle began in 2008.

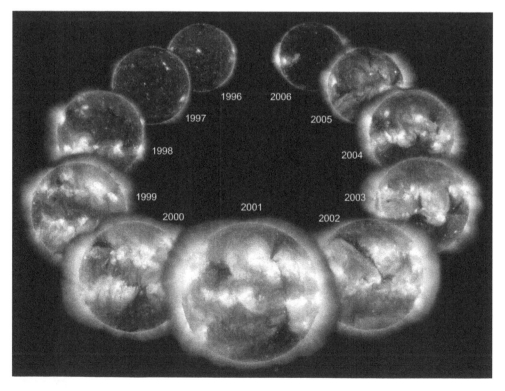

Figure 4.11. Nearly an entire solar activity cycle imaged in ultraviolet light from the SOHO spacecraft.

Why is it important to keep track of the sunspot cycle? _____

_____

_____

Answer: The Sun is most active during the years of sunspot maximums, pouring the great-est amount of energy and radiation into Earth's environment.

4.11 MAGNETISM

Sunspots are like huge magnets. These regions of powerful magnetic fields are typically thousands of times stronger than Earth's magnetic field.

The magnetic field of a sunspot can be detected before the spot itself can be seen and after the spot is gone. Therefore, magnetic fields probably shape and control local conditions on the Sun. Astronomers analyze magnetic fields by measuring Zeeman spectral line splitting (Section 3.10).

A weaker magnetic field spreads out over the whole Sun. It has a north magnetic pole and a south magnetic pole, with the magnetic axis tilted 15° to the rotation axis. It is split into two hemispheres. A display of magnetic field strength is called a **magnetograph**.

The Sun's magnetic field extends from its northern hemisphere through the solar system out at least 16 billion km (10 billion miles). Near the edge of the solar system, the magnetic field bends and returns to the Sun's southern hemisphere.

The complex solar magnetic field is generated by rotational and convec-tive motions of electrically charged particles that make up the Sun's hot gases. Apparently it energizes and controls violent outbursts of material and radia-tion on the Sun.

The polarity of the Sun's magnetic field is reversed about every 11 years, shortly after the period of sunspot maximum. It takes two sunspot cycles of about 11 years each for the Sun's magnetic poles and sunspot magnetic polari-ties to repeat themselves. So the solar activity cycle is 22 years when counting the length of time required for the Sun to return to its original configuration.

What probably activates the violent outbursts of material that occur on the Sun? _____

Answer: Very strong magnetic fields at the sites of sunspots.

You can observe a magnetic field by putting a magnet under a piece of paper. Lightly sprinkle iron filings on top of the paper. The filings will line up according to the strength of the magnetic force. By showing the regions of the magnetic force, they make the magnetic field visible to you.

4.12 FLARES AND CORONAL MASS EJECTIONS

A solar **flare** is a sudden, tremendous, explosive outburst of light, invisible radiation, and material from the Sun. One great solar flare may release as much energy as the whole world uses in 100,000 years.

Flares are short-lived, typically lasting a few minutes. The largest last a few hours. They occur near sunspots, especially in periods of sunspot maximums. Flares seem to be energized by strong local magnetic fields.

A flare usually follows the most energetic of all solar eruptions, a coronal mass ejection (CME), a large burst of mass that expands as it climbs. A typical

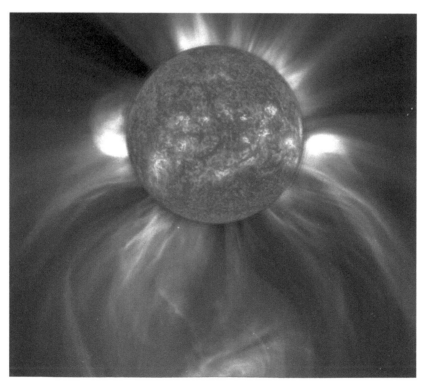

Figure 4.12. A widely spreading coronal mass ejection blasts more than a billion tons of matter out into space at millions of km an hour. Image from a SOHO spacecraft coronagraph. An enlarged ultraviolet image from a different day fills the occulting disk for effect.

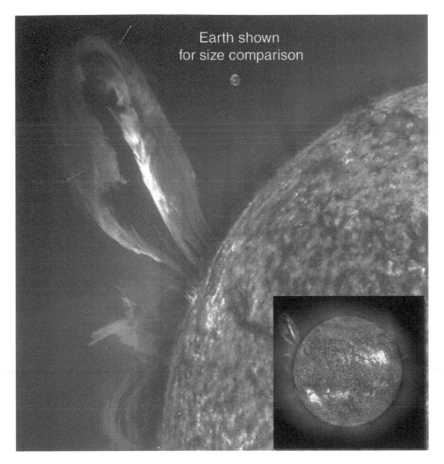

Figure 4.13. Ultraviolet image of an erupting prominence that extends over 35 Earths out from the Sun. Dot represents Earth for size comparison.

CME blasts billions of tons of plasma out from the corona (Figure 4.12). The mass moves out into the solar system along magnetic field lines at speeds of over 600 km/sec (1 million miles/hour).

U.S. robot Interface Region Imaging Spectrograph (IRIS) (2013–) is exploring the region between the photosphere and corona in order to understand the source of the heat that drives material.

What probably bends and controls the trajectory of the ejected gases in solar flares, coronal mass ejections, and prominences?_____

_____

Answer: Strong magnetic fields in the vicinity of sunspots.

4.13 HOW SOLAR ERUPTIONS AFFECT EARTH

Huge flares and coronal mass ejections hurl fantastic amounts of high-energy radiation and electrically charged particles—as much energy as a billion exploding hydrogen bombs—into the solar system.

Gamma rays, X-rays, and ultraviolet rays reach Earth in just 8.3 minutes. Charged particles arrive a few hours or days later. These could destroy all life if Earth were not shielded by its magnetic field and atmosphere. Extra solar radiation is risky for airplane passengers, astronauts, and spacecraft electronics.

When electrically charged particles from the Sun strike Earth's atmosphere, they can stimulate the atmospheric atoms and ions to radiate light, producing **aurora borealis** (northern lights) and **aurora australis** (southern

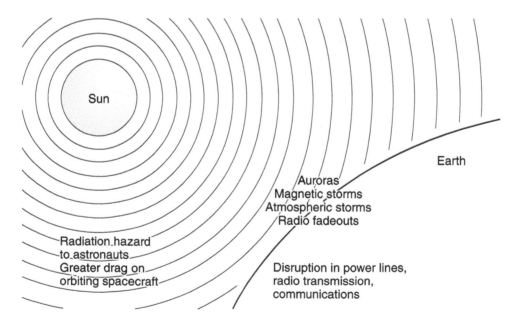

Figure 4.14. Effects of solar flares and coronal mass ejections on Earth's environment.

lights). Auroras are bands of light seen in the night sky in the Arctic and Antarctic regions but occasionally also down to middle latitudes about 2 days after a solar flare. They reach a peak about 2 years after sunspot maximum.

Strong blasts of electric solar particles interact with Earth's magnetic field and disturb it, causing geomagnetic storms. Compasses don't work normally then. The gusts can cause atmospheric storms, satellite damage, surges in electric power and telephone lines, and blackouts.

High-energy radiation heats the upper atmosphere, making it expand. Then friction and the drag on spacecraft in low Earth orbits increase. The drag is greatest during times of maximum solar activity, when satellites may plunge from orbit and be destroyed on reentry. The first U.S. space station, Skylab (1973–1979), was a casualty of a solar maximum. By increasing ionization, solar outbursts can disrupt radio transmission and navigation signals.

Because solar flares affect modern life so much, the U.S. National Oceanic and Atmospheric Association (NOAA), with partners worldwide, monitors the Sun's magnetic field and activity daily (Figure 4.14). Weather Prediction Center ▶www.swpc.noaa.gov◀ issues space weather alerts, warnings, and forecasts.

List two effects that large solar outbursts have on modern technology on Earth. (1) _____ ; (2) _____

_____

Answer: (1) Disruption in power transmission; (2) disruption in communications.

4.14 SOLAR WIND

The **solar wind** is a plasma, or stream of energetic, electrically charged particles that flows out from the Sun at all times. It is much faster, thinner, and hotter than any wind on Earth.

The solar wind is observed by instruments carried on spacecraft above Earth's atmosphere. Near Earth, the average solar wind speed is about 450 km/second (1 million miles/hour). Earth's atmosphere and magnetic field ordinarily protect us from harmful effects of the solar wind.

Big blasts of solar wind occur at coronal mass ejections. The wind is strongest during periods when many sunspots are visible and solar activity is great. Strong blasts of solar wind produce especially brilliant auroras. ▶www.spaceweather.com◀

The solar wind comes mainly from **coronal holes**, regions in the Sun's corona where gases are much less dense than elsewhere. Magnetic fields are relatively weak there, allowing high-speed solar wind streams to escape.

Voyager spacecraft instruments (Section 8.12) continue to measure the solar wind in the heliosheath, the region where the solar wind is slowed to subsonic speeds by the interstellar gas pressure.

What is the solar wind? _____

_____

Answer: A stream of energetic, electrically charged particles that flows out from the Sun.

4.15 PROBING THE INTERIOR

Scientists believed that they understood what makes the Sun shine until **solar neutrino experiments** raised some doubts.

Theoretically, the Sun's energy is produced by the conversion of hydrogen into helium in nuclear fusion reactions. **Solar neutrinos**, neutral elementary particles with almost no mass, are a by-product. Neutrinos interact very weakly with matter and travel at virtually the speed of light.

Scientists cannot look directly deep inside the Sun's core to test their theory. But the theory predicts that the neutrinos produced in the core should escape. So they look for the solar neutrinos instead.

Neutrinos produced in the center of the Sun have been detected. They show that the theory is basically correct.

Scientists built several neutrino traps deep inside the Earth. The number of neutrinos detected over 30 years was lower than the number theory predicts. New experiments and analysis explain that this **solar neutrino problem** was due to neutrino oscillations. Neutrinos change from one type to another, and some were not counted before.

Helioseismology is the study of the Sun's internal structure and condition by measuring global oscillations on its surface. Pressure waves of different frequencies penetrate to different depths. They reveal the density, temperature, and rotation rate inside the Sun, just as seismic waves expose Earth's interior. Solar oscillations are observed spectroscopically through Doppler shifts in spectral lines (Section 3.9). Six stations of Global Oscillation Network Group (GONG) are located around the world to obtain nearly continuous observations. ▶www.gong.noao.edu◀

Explain the small number of solar neutrinos detected in early experiments.

_____

Answer: Neutrino oscillations: Neutrinos change from one type to another, and some were not counted before. (Astronomers rely on the results of the experiments.)

4.16 COMMON FEATURES

Astroseismology extends this study to other stars. Apparently other stars have regions of violent activity like those of our Sun, including **starspots** and **starspot cycles**, although stars are so far away that these must be deduced from their spectral lines and brightness variations rather than observed directly. Recent X-ray observations indicate that nearly all types of stars also have similar coronas with temperatures of at least a million degrees.

Write a short summary describing three phenomena that indicate violent activity on the Sun (and by extension to other stars), and name their probable cause.

_____

_____

_____

Answer: Your answer should briefly describe (1) sunspots, or dark, relatively cool, temporary spots on the Sun's photosphere; (2) flares, or sudden, short-lived outbursts of light and material near a sunspot; (3) coronal mass ejections, or blasts of plasma from the corona.
Probable cause: the Sun's powerful magnetic fields.

Most violent activity on the Sun seems to be caused and controlled by very strong local magnetic fields.

4.17 MOTIONS IN SPACE

The Sun, like all other stars, is racing through space.

With respect to nearby stars, the Sun is speeding toward the constellation Hercules at 20 km/second (45,000 miles/hour), carrying its planets along with it.

The Sun with its planets is inside the Milky Way Galaxy. It goes around our Galaxy's center as the whole Galaxy turns around in space. The Sun travels at about 250 km/second (563,000 miles/hour) (Section 6.2).

SELF-TEST

This self-test is designed to show you whether or not you have mastered the material in Chapter 4. Answer each question to the best of your ability. Correct answers and review instructions are given at the end of the test.

1. List three reasons why modern astronomers study the Sun.

   (1)_____

   _____

   _____

   (2)_____

   _____

   (3)_____

   _____

   _____

2. Match the most appropriate tool to the work:

   ____ (a) Image processes in the hottest active regions of the Sun.

   ____ (b) Image corona outside solar eclipse.

   ____ (c) Photograph the Sun's visible surface.

   ____ (d) Image the Sun in the light of a particular element.

   ____ (e) Receive and record solar radio waves.

   (1) Coronagraph.
   (2) Optical solar telescope.
   (3) Radio telescopes.
   (4) Spectroheliograph.
   (5) Ultraviolet ray, X-ray, and gamma ray telescopes.

3. Define the astronomical unit (AU). _____

   _____

4. Sketch the Sun, and identify the corona, chromosphere, photosphere, convection zone, radiation zone, and core.

5. Estimate (a) diameter; (b) mass; and (c) surface temperature of the Sun.

   (a) _____ ; (b) _____ ; (c) _____

6. Why is the sunspot cycle carefully monitored from Earth? _____

_____

_____

_____

7. Identify the following phenomena of the Sun:

____ (a) Most energetic eruption of material from the solar corona.

____ (b) Bright cell that looks like rice grain in the photosphere.

____ (c) Dark, relatively cool blotches in the bright photosphere.

____ (d) Elementary particles predicted to be produced in nuclear reactions in the core.

____ (e) Tremendous, short-lived, explosive outburst of light and material.

(1) Flare.

(2) Granule.

(3) Coronal mass ejection.

(4) Solar neutrino.

(5) Sunspot.

8. What is the solar wind? _____

_____

9. List four ways that a flare and unusually big blasts of solar wind can affect Earth's environment. (1) _____

(2) _____

(3) _____

(4) _____

10. (a) What is the solar constant? (b) Why is it important to know if it is truly constant or if it varies with time? _____

_____

_____

_____

_____

ANSWERS

Compare your answers to the questions on the self-test with the answers given below. If all of your answers are correct, you are ready to go on to the next chapter. If you missed any questions, review the sections indicated in parentheses following the answer. If you missed several questions, you should probably reread the entire chapter carefully.

1. (1) The Sun is a free, nonpolluting, almost inexhaustible source of present and potential future energy.

   (2) The Sun is the only star close enough to observe in detail, so astronomers use it to determine what other stars are like.

   (3) Changes in the Sun's energy output affect Earth's climate, atmosphere, and weather, as well as power-transmission and communications systems.   (Sections 4.1, 4.13)

2. (a) 5;   (b) 1;   (c) 2;   (d) 4;   (e) 3.   (Section 4.7)

3. The astronomical unit (AU) is the average distance between the Earth and the Sun, about 150 million km (93 million miles) (officially 149,597,870 km).   (Section 4.2)

4. See Figure 4.6, Regions of the Sun. (a) Corona;   (b) chromosphere;   (c) photosphere;   (d) convection zone;   (e) radiation zone;   (f) core.   (Section 4.4)

5. (a) 1,390,000 km (864,000 miles);   (b) 2 x 10^{30} kg;   (c) 5800 K (10,000° F).   (Sections 4.2, 4.4, 4.6)

6. The sunspot cycle is watched carefully from Earth as an indicator of solar activity. The Sun is most active, with greatest outbursts of energy and radiation, during the years when sunspots are most numerous. It is least active in the years of sunspot minimums.   (Sections 4.10, 4.13, 4.14)

7. (a) 3;   (b) 2;   (c) 5;   (d) 4;   (e) 1.   (Sections 4.8, 4.9, 4.12, 4.14, 4.15)

8. A stream of energetic, electrically charged particles that flows out from the Sun.   (Section 4.14)

9. (1) Increased hazardous radiation;   (2) auroras;   (3) magnetic storms;   (4) atmospheric storms.   (Sections 4.13, 4.14)

10. (a) The amount of the Sun's energy that falls per second on Earth's outer atmosphere, about 1400 watts per square meter (126 watts per square foot).   (b) Changes in the solar constant might drastically change Earth's climate and atmosphere.   (Sections 4.1, 4.13)

5

STELLAR EVOLUTION

To every thing there is a season, and a
time to every purpose under the heaven.
A time to be born, and a time to die.

Ecclesiastes 3:1–2

Objectives

☆ Define stellar evolution.

☆ List the stages in the life cycle of a star like our Sun according to the modern theory of stellar evolution.

☆ Explain the importance of the H–R diagram to theories of stellar evolution.

☆ Explain the relation between a star's age and its position on the H–R diagram.

☆ List the main steps in the birth of a star.

☆ Describe the energy balance and pressure balance in main sequence stars.

☆ Compare and contrast what happens in the advanced stages of evolution for stars of large and small mass: planetary nebulas, white dwarfs, supernovas, pulsars/neutron stars, and black holes.

☆ Identify nebulas, main sequence, blue giant, red giant, and pulsating variable stars that can be observed in the sky.

☆ Explain how supernovas and pulsars are observed.

☆ Describe the origins of the different chemical elements and the importance of supernovas to new generations of stars.

☆ Describe observational evidence for black holes.

5.1 LIFE CYCLE OF STARS

No star shines forever. **Stellar evolution** refers to the changes that take place in stars as they age—the life cycle of stars. These changes cannot be observed directly, because they take place over millions or billions of years. Astronomers construct a theory of stellar evolution that is consistent with the laws of physics. Then they check their theory by observing real stars shining in the sky.

In checking theory against observations, astronomers make use of H–R diagrams. Theoretical predictions are made regarding a sequence of changes in luminosity and temperature for stars as they go from birth to death. These changes are plotted on an H–R diagram, forming theoretical **tracks of evolution**. Theoretical H–R diagrams are then compared with H–R diagrams for groups of real stars (Section 6.4).

The predictions of the **modern theory of stellar evolution**, described in this chapter, agree well with the data from observations of real stars.

What is stellar evolution?_____

_____

Answer: The changes that take place in stars as they age—the life cycle of stars.

5.2 BIRTHPLACES

Stars form out of matter that exists in space. The gigantic **interstellar** (between stars) **clouds** of gas and dust must be the birthplaces of stars.

You can see the nearest cloud in space where new stars are forming now. The famous Orion Nebula, located about 1500 light-years away in the constellation Orion, is a region of intense star formation (Figure 5.1).

Look for the Orion Nebula in the winter. It is marked on your winter skies map in the sword of Orion the Hunter. The Orion Nebula looks like a hazy patch to your eye. Through a telescope you will see it glow with a greenish color. Hot, newly formed stars in the region make the gases glow. A much larger associated dark molecular cloud is not visible.

Are new stars still being born today? Where? _____

_____

Answer: Yes. In gigantic clouds of gas and dust, as in the Orion Nebula.

Figure 5.1. (a) Orion Nebula, in the constellation Orion. (b) Orion the Hunter.

5.3 BIRTH

A **protostar** is a star in its earliest observable phase of evolution. You can think of a protostar as a star that is being born.

Protostars form by chance at high-density spots inside huge, cool, turbulent gas (mostly hydrogen) and dust clouds that exist in space. Perhaps a shock wave from an exploding star (supernova) triggers the process.

Initially, the force of gravity pulls matter in toward the center of a dense clump, causing it to contract and become even denser. Matter continues to accrete onto the shrinking protostar and surrounding disk. Energetic jets and winds rise out from both poles, disrupting the birth cloud. Gravitational contraction causes the temperature and pressure inside the protostar to rise greatly.

Heat flows from the protostar's hot center to its cooler surface and encircling accretion disk. The protostar radiates this energy, and the disk reradiates it also. They shine at infrared, millimeter, and submillimeter wavelengths.

Particles in the disk may accrete to form planets (Figure 12.2). At this stage, the newly forming star is called a young stellar object (YSO).

When the temperature in the YSO's center reaches 10 million K, nuclear fusion reactions start. These nuclear reactions release tremendous amounts of energy. Energy is generated in the center as fast as it is being radiated out into space. The very high internal temperatures and pressures are thus maintained.

The outward pressure of the very hot gases balances the inward pull of gravity (Figure 5.2). This balance is called **hydrostatic equilibrium**. The YSO stops contracting. It shines its own light steadily into space. The YSO

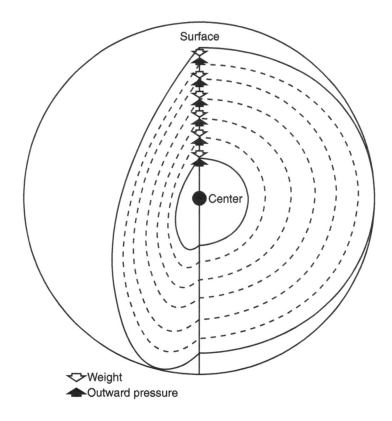

Figure 5.2. The outward gas pressure balances gravity at every level in a star.

becomes a newborn star. Most likely our Sun was born in this way about 5 billion years ago.

Recent observations support this theory of star birth. Protostars in dense cores of gaseous clouds, circumstellar disks, and jets of gas streaming away from YSOs are imaged at infrared, millimeter, and submillimeter wavelengths.

List the three main steps in the birth of a star. (1) _____

_____ ;

(2) _____ ;

(3) _____

Answer: (1) Gravitational contraction within a cloud of gas and dust; (2) rise in interior temperature and pressure; (3) nuclear fusion.

5.4 LIFETIMES

The clouds in which protostars form do not have identical masses or distributions of the chemical elements. The life cycle of a star—the time it takes for a star to evolve—depends upon its initial **mass** and **chemical composition**.

Stars that begin life with about the same mass and chemistry go through the same stages of evolution in about the same amount of time.

Stars of similar chemistry with very high mass evolve fastest, while those of very low mass take the longest time to evolve.

The theoretical evolutionary tracks on the H–R diagram in Figure 5.3 show how a protostar's luminosity and temperature change as it contracts to become a star.

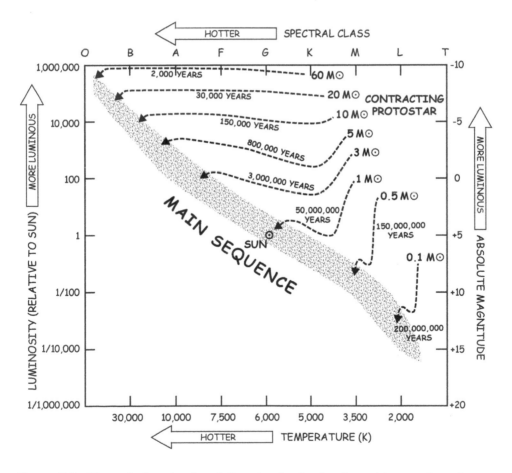

Figure 5.3. Theoretical tracks of evolution showing luminosity and temperature changes of contracting protostars of different masses. (Contraction times are marked on the tracks.)

Approximately how long does it take each of the following protostars to reach zero age main sequence (to be born)? (a) stars like our Sun _____ _____ ; (b) stars with mass much greater than the Sun's _____ ; (c) stars with mass much less than the Sun's _____

Answer: (a) About 50 million years; (b) about 2000 years; (c) about 200 million years.

5.5 WHY STARS SHINE

You can think of a **main sequence** star as an adult star. In comparison to changes in protostars, evolution of main sequence stars is very slow. A star spends most of its lifetime shining steadily, with luminosity and temperature values found along the main sequence of H–R diagrams.

A main sequence star gets its energy from **nuclear fusion reactions** in which hydrogen at the center of the star is converted into helium (Figure 5.4). Four hydrogen nuclei are fused into one lighter helium nucleus. The disappearing mass is changed into energy and released. (The same process releases energy in hydrogen bombs.)

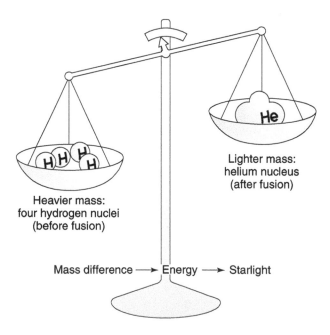

Figure 5.4. An imaginary experiment showing why the stars shine. If you could weigh the hydrogen nuclei before and the helium nucleus after fusion, you would discover that the helium nucleus was lighter.

The energy from the nuclear fusion reactions eventually reaches the star's surface. Then the star shines energy into space.

The amount of energy released in a nuclear fusion reaction can be calculated from the famous special relativity theory equation of (German-born) U.S. physicist Albert Einstein:

$$E = mc^2$$

where E = energy, m = mass difference, and c = speed of light.

According to Einstein's equation, when many nuclear fusion reactions occur together, enormous amounts of energy are released. The Sun is a huge, hot gaseous sphere that shines steadily without appreciable change of size or temperature. Although practically 5 million tons of hydrogen must be converted into helium each second to produce the Sun's luminosity, less than 0.01 percent of the Sun's total mass changes to sunshine in a billion years.

What is the source of energy that lets main sequence stars shine?

_____

Answer: Nuclear fusion reactions in which hydrogen is converted into helium.

5.6 OLD AGE ★

A star will shine steadily as a main sequence star until all the available hydrogen in its core has been converted into helium. Then the star will begin to die.

Our Sun is an average medium-sized star. It has been shining as a stable main sequence star for about 5 billion years, and it should continue to shine steadily for another 5 billion years.

Very massive, hot, bright stars die fastest because they use up their hydrogen most rapidly. The very massive blue giant stars, such as Rigel in Orion, spend only a few million years shining as main sequence stars.

The least massive, cool, dim stars live the longest because they consume their hydrogen fuel least rapidly. The small-mass red dwarfs are the oldest and most numerous main sequence stars. They have lifetimes billions of years long.

What types of stars are expected to live (a) longest? _____

_____ (b) shortest? _____ (c) About how

much longer is the Sun expected to shine as it does now? _____

_____

_____

Answer: (a) Those with small mass, such as red dwarfs; (b) very massive stars, such as blue giants; (c) about 5 billion years.

5.7 RED GIANTS

After the hydrogen fuel in the star's core is used up, the star no longer has an energy source there. The core, which then consists primarily of helium, begins to contract gravitationally. Hydrogen fusion continues in a shell around the helium core, under the outside envelope of hydrogen.

Gravitational contraction causes the temperature of the helium core to rise. The high temperature makes the shell hydrogen fuse faster, and the star's luminosity increases.

The tremendous energy released by this hydrogen fusion and gravitational contraction heats up surrounding layers. The star expands to gigantic proportions. The star's density is then very low everywhere except in the core (Figure 5.5).

As the star expands, its surface temperature drops and its surface color turns to red. The star has changed into a huge, bright, red, aging star—a **red giant** or **supergiant**. It is cool but bright because of its gigantic surface area. It has the luminosity and temperature values of the upper-right region of the H–R diagram.

You can see some red supergiant stars shining in the sky. Good examples are Betelgeuse in Orion and Antares in Scorpius, both over 400 times the Sun's diameter (Tables 1.1 and 2.1).

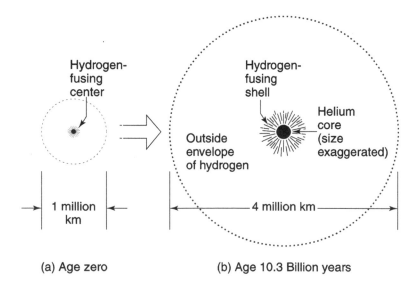

Figure 5.5. Sunlike star (a) at start of life on main sequence and (b) as it ages to red giant.

Our Sun, like all stars, is expected to change into a huge red giant when it dies. That red giant Sun will shine so brightly that rocks will melt, oceans will evaporate, and life as we know it on Earth will end.

When does a star begin to change from a main sequence star into a red giant?

Answer: When it has converted all of the available hydrogen fuel in its core into helium.

5.8 SYNTHESIS OF HEAVIER ELEMENTS

Gravitational contraction causes the temperature inside the red giant's helium core to rise to 100 million K. At that temperature, helium is converted to carbon in nuclear fusion reactions (Figure 5.6).

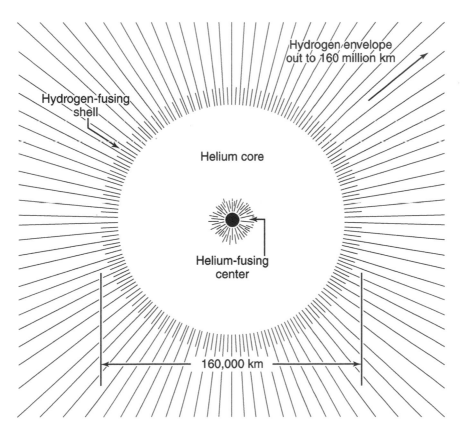

Figure 5.6. Structure of a red giant star.

The helium core does not expand much once the helium fusion starts. The temperature builds up rapidly without a cooling, stabilizing expansion. Helium nuclei fuse faster and faster, and the core gets even hotter. This nearly explosive ignition of helium fusion is called the **helium flash**.

After some time, the temperature rises sufficiently so that the core expands. Cooling occurs inside, and helium fusion goes on at a steady rate, surrounded by a hydrogen-fusing shell.

Inside the more massive red giants, further fusion reactions can build up familiar elements heavier than carbon, such as oxygen, aluminum, and calcium (Appendix 4).

Astronomers believe that elements like carbon and oxygen, which we need for life, are made where? _____

Answer: Inside red giant stars.

5.9 VARIABLE STARS ★

A star moves back and forth in the area between the red giant region and the main sequence several times, in a way not yet fully understood, before it enters the final stages of its life.

Most stars probably change from red giants to **pulsating variable stars** before they finally die. That is, they expand and contract and grow bright and fade periodically.

Cepheid variables are very large luminous yellow stars whose light output varies in periods from 1 to 70 days. You can observe Delta Cephei, the first discovered and the star for which this class of variables was named (Figure 5.7). Cepheids are important because they provide a way of measuring distances too great to be measured by trigonometric parallax.

More than 700 Cepheid variables are known in the Milky Way Galaxy. Polaris, the North Star, is the nearest. Its brightness varies between magnitudes 2.5 and 2.6 about every 4 days.

U.S. astronomer Henrietta Leavitt (1868–1921) discovered that the longer the period of light variation of Cepheids, the greater the luminosity. Astronomers use this **period-luminosity relation** to determine the absolute magnitude of Cepheids after measuring their periods.

A comparison of the calculated absolute magnitude and the observed apparent magnitude yields the distance to the Cepheids and the star groups they belong to (Section 3.16). Cepheids are useful **distance markers** out to about 3 Mpc (10 million light-years).

RR Lyrae variables, named after variable star RR in the constellation Lyra, are pulsating blue-white giants whose light output varies from brightest to

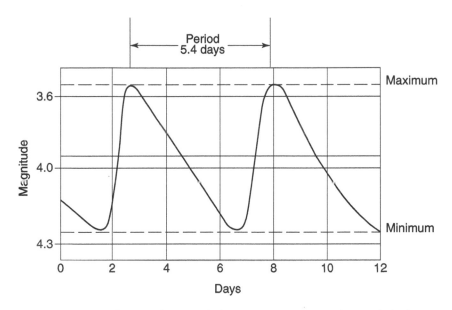

Figure 5.7. Light curve showing how the light output varies for Delta Cephei, the prototype Cepheid variable star.

dimmest in periods of less than a day. About 4500 RR Lyrae stars are known in the Milky Way Galaxy. RR Lyrae stars are used to measure the distance to the star clusters they belong to, out to about 200,000 pc (600,000 light-years).

Long-period Mira variables, named for famous Mira in the constellation Cetus, are red giants that take between 80 and 1000 days to vary between brightest and faintest. Mira, about 40 pc (130 light-years) away, varies from its maximum bright red to its minimum output, where it becomes invisible, in a period of 332 days. Mira was named the "Wonderful" by amazed seventeenth-century observers, who first recorded its brightness fluctuations.

Name three characteristics of a pulsating variable star that change periodically. (1) _____ ; (2) _____ ; (3) _____

Answer: (1) Size; (2) luminosity; (3) temperature.

5.10 DEATH

All stars evolve in about the same way, although over different periods of time, until their cores become mostly accumulated carbon (Figure 5.8). The last stage in a star's evolution, or the way it finally dies, depends greatly on its mass.

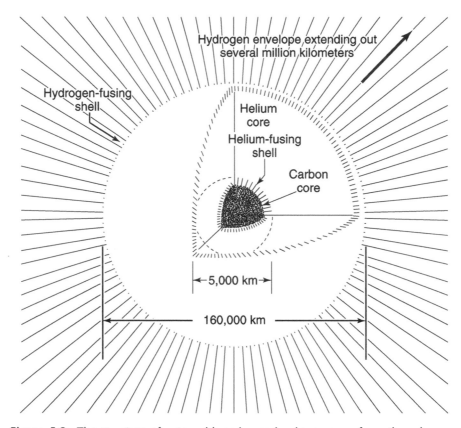

Figure 5.8. The structure of a star with an increasing inner core of mostly carbon.

Small stars, up to about 1.4 times the Sun's mass, finally die without a fuss, quietly fading away into the blackness of space. Very massive stars end with a violent explosion, flaring up brilliantly before giving up life.

What characteristic of a star determines the way it finally dies? _____

Answer: Its mass.

5.11 MASS LOSS

When a star of mass like our Sun has depleted all of its available helium fuel, it becomes a bloated red giant star for the last time. (At this stage of its life our Sun will become so big that it will swallow up Mercury, Venus, Earth, and Mars.)

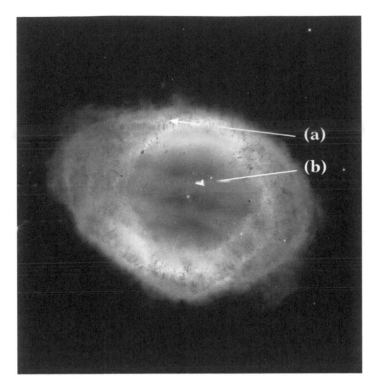

Figure 5.9. Famous Ring Nebula in Lyra, a planetary nebula and star core.

The star then throws off some of its mass. The star's outermost hydrogen envelope, enriched by heavier elements, flies off into space. Electrically charged particles stream away in a flow called a **stellar wind**. (The solar wind is described in Section 4.14.) Deeper layers are thrown off in a wispy, expanding shell of gas typically about 0.5 to 1 light-year across, called a **planetary nebula**, which continues to spread out at speeds of about 20 to 30 km/sec (45,000 to 67,500 miles/hour). The star's core is left behind.

About 1600 planetary nebulas have been recorded. They are probably less than 50,000 years old, because the gas atoms in the nebula separate rapidly. After about 100,000 years, the shell is too spread out to be visible.

Examine Figure 5.9. Identify the core of the star and the planetary nebula in the photograph. (a) _____ ;
(b) _____

Answer: (a) Planetary nebula; (b) core of the star.

5.12 WHITE DWARFS

After it has thrown off its gas envelope, the star remains as a core of carbon surrounded by a shell of burning helium.

A star that has exhausted all of its nuclear fuel can no longer withstand the pull of gravity. It contracts again as gravity pulls matter in toward the center. Gravitational contraction makes the temperature and pressure go up very high, and electrons are stripped off atoms. The star becomes a small, hot, **white dwarf**. It is made mostly of electrons and nuclei. These subatomic particles can be squeezed much closer together than whole atoms can.

Eventually, when the white dwarf star reaches about Earth's size, it cannot contract any further. White dwarf stars of mass like the Sun are very dense because gravity packs all that mass into a star the size of Earth. The force of gravity on such a white dwarf star would be about 350,000 times greater than that on Earth. If you could stand on a white dwarf star, you would weigh 350,000 times more than you do on Earth.

If a white dwarf in a binary system accretes matter from its companion star, it may briefly blaze, called a **nova**. Or, it may explode brilliantly when a wave of nuclear fusion rips through a bigger buildup, called a **Type Ia supernova**.

Usually a white dwarf star cools, turns to dull red, and shines its last energy into space. Then the white dwarf becomes a dead **black dwarf** in the graveyard of space.

What is a white dwarf star?_____

_____

_____

Answer: A small dense (dying) star of low luminosity and high surface temperature, typically about the size of Earth but with mass equal to the Sun's.

5.13 LIFE CYCLE OF SUNLIKE STARS

Identify each stage of the life of a star like our Sun, as labeled sequentially in Figure 5.10. (a) _____ ;

(b) _____ ;

(c) _____ ;

(d) _____ ;

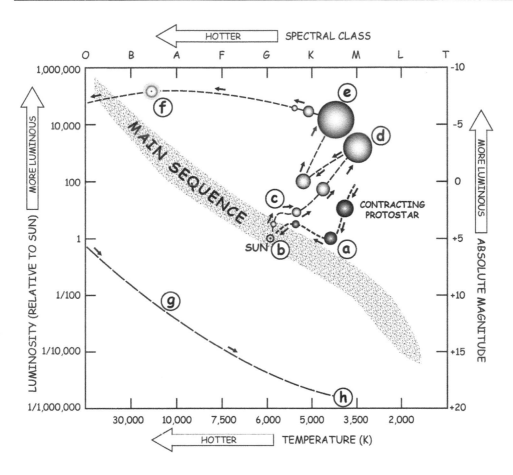

Figure 5.10. The life stages of a star like our Sun.

(e) _____ ;

(f) _____ ;

(g) _____ ;

(h) _____

Answer: (a) Protostar, gravitational contraction of cloud of gas and dust; (b) stable main sequence star, shining by nuclear fusion (converting hydrogen to helium); (c) evolution to red giant when helium core forms; (d) red giant, shining by helium fusion; (e) variable star, formation of carbon core; (f) planetary nebula, enriched hydrogen envelope ejected into space; (g) white dwarf, mass packed into star about the size of Earth; (h) dead black dwarf in space.

5.14 EXPLODING STARS

A **supernova** is a gigantic stellar explosion. It may outshine its whole galaxy for a short time.

Most stars of eight or more times the Sun's mass die in a spectacular explosion called a **Type I sp e n u** . Their carbon cores contract gravitationally in the same way that a smaller star's does. But in a red giant star, the core temperature continues to rise to 600 million K. Then the carbon fuses into magnesium. The collapse stops when the carbon in the core is used up. A new cycle begins—core contraction with rising temperature and onset of new nuclear reactions, fusion of heavier elements such as oxygen and silicon, new shells of lighter elements, and a halt in the collapse.

Iron ends these cycles, because it requires rather than releases energy in nuclear reactions. The doomed core collapses for the last time. When it cannot be compressed any further, it rebounds, sending out a shock wave. The outer layers explode violently. Light from the supernova can reach 100 billion times the Sun's luminosity.

Most of the energy released in the explosion is invisible. A great amount is carried away at the speed of light by high-energy radiation and neutrinos ejected from the collapsing core. This energy holds clues to the causes of stellar explosions and the kinds and amounts of chemical elements manufactured and sprayed into space by supernovas.

Supernova 1987A, the first bright supernova in the sky since 1604, appeared in neighbor galaxy the Large Magellanic Cloud in 1987. It was visible from the southern hemisphere for months and is the best-observed supernova to date (Figure 5.11). Neutrinos were detected exactly as theory predicted. The core temperature at explosion must have been 200 billion K! Now astronomers are using Supernova 1987A data to refine and test theories of star death.

What kind of stars die as Type II supernovas? _____

Answer: Very massive stars (about 8 or more times the Sun's mass).

5.15 SUPERNOVA REMNANTS

You might say that you are made of star dust.

Hydrogen and helium were probably the only elements in the universe when it began. Elements such as carbon, oxygen, and nitrogen, essential for life, are made inside the fiery cores of aging stars. The heaviest elements of all,

Figure 5.11. Supernova 1987A (top) in 1969 before the explosion was seen and (bottom, right of center) a week afterward, in February 1987.

such as gold and lead, are produced in the extremely high temperatures and intense neutron flux of a supernova explosion.

The supernova explosion sprays all these new elements out into space. They mix with the hydrogen, helium, and dust already there. All the material scattered into space by exploding massive stars becomes available again to be used in the formation of new stars and planets. Our Sun and Earth were formed about 4.6 billion years ago from a cloud of hydrogen and helium enriched in this way.

In A.D. 1054, Chinese and Native American observers recorded seeing a brilliant new star blaze in the sky even during daylight hours. The Crab Nebula in Taurus, a gas cloud expanding at 1600 km (1000 miles) per second, is observed today at the site of that supernova. It is about 3 pc (10 light-years) across, with the remnant core of the exploded star still at the center (Figure 5.12).

Which do you think are more abundant in the universe, elements lighter than iron or those heavier than iron? Why? _____

_____

_____

_____

_____

Answer: Lighter elements. These elements have much more time to form. Elements lighter than iron are produced from primordial hydrogen over a long period of time inside the cores of massive stars, while those that are heavier are produced only during the brief interval when the star explodes (supernova) at the end of its life.

5.16 SUPERDENSE STARS

When a very massive star explodes, it may leave behind a star of more mass than the Sun squeezed tightly together into a ball only about 16 km (10 miles) across. This extremely dense star is made mostly of **neutrons**, uncharged atomic particles. It was named a **neutron star** when it was first hypothesized.

Pulsars, pulsating radio stars, were first observed in 1967 by Jocelyn Bell, a graduate student at Cambridge University, England. Pulsars send sharp, strong bursts of radio waves to Earth with clocklike regularity, at intervals between milliseconds and 4 seconds. **Magb as** have abnormally strong magnetic fields, which lead to X-ray and gamma ray emission.

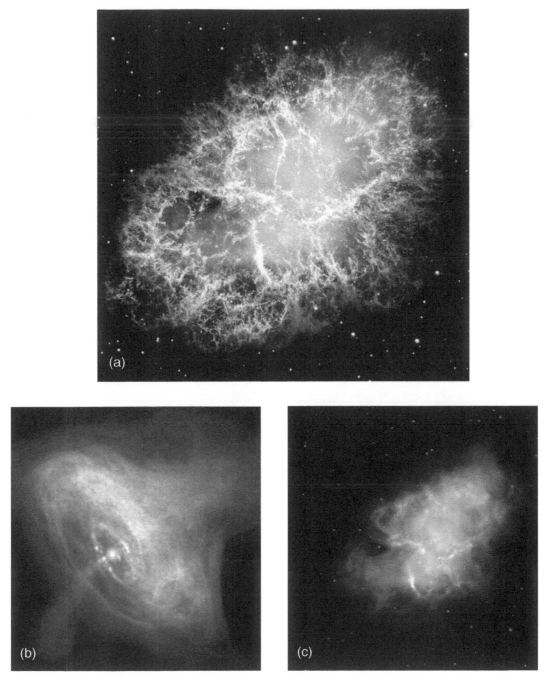

Figure 5.12. The Crab Nebula (M1) and pulsar (white dot near center) in Taurus are bright sources of radiation in all wavelengths. (a) Optical image. (b) X-ray image. (c) Infrared image.

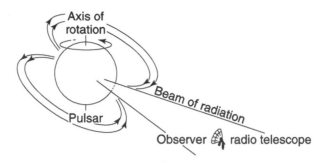

Figure 5.13. A pulsar or neutron star. Astronomers observe regular pulses of radiation emerging from the rotating star's magnetic poles as they sweep past Earth.

Theory predicted that a neutron star should exist at the center of the Crab Nebula. A pulsar was found there in 1968 (Figure 5.12). The Crab Pulsar has since been observed over all electromagnetic wavelengths, from radio to gamma.

A pulsar is a rapidly rotating, highly magnetic neutron star (Figure 5.13). Its characteristic short, regular pulses come from radiation beams, emitted by very energetic accelerated charged particles, sweeping past Earth as the neutron star periodically spins. The rotation and pulse rates gradually slow down as energy is radiated away.

X-ray bursters blast X-rays randomly. X-ray bursts come from an accreting neutron star in a binary system when its big, hot helium buildup explodes.

How would you expect the force of gravity on the surface of a pulsar to compare to the force of gravity on Earth?

_____

_____

Answer: Much greater on a pulsar. The force of gravity is stronger the closer matter is packed, and a pulsar is extremely dense.

5.17 BLACK HOLES

A really massive star may continue to collapse after the pulsar stage to become a bizarre object called a **black hole**.

Figure 5.14. (a) An optical image that spans 700 light-years shows black hole Cygnus X-1, outlined in a box. Art depicts what astronomers think is happening within the Cygnus X-1 system.

If black holes do exist, they are not holes at all. On the contrary, a black hole is a large mass contracted to extremely small size and enormous density. The force of gravity in such an object would be so great that, according to Einstein's theory of relativity, it would suck in all nearby matter and light.

A black hole can never be seen, because no light, matter, or signal of any kind can ever escape from its gravitational pull—hence its name. The surface of a black hole, or the boundary through which no light can get out, is called the **event horizon**.

The **Schwarzschild radius (R_S)** is the critical radius at which a spherically symmetric massive body becomes a black hole. The equation is:

$$R_S = 2GM/c^2$$

where G is the gravitation constant, M is the mass of the body, and c is the speed of light (Appendix 2). The Schwarzschild radius for the Sun is about 3 km (2 miles) while for Earth it is about 1 cm (0.4 inch).

Theory predicts that a star of over three solar masses at its final collapse must cross its event horizon and disappear from view. No known force could stop further collapse, so the star may continue to shrink to a spot at the center called a **singularity**.

No one can observe black holes directly. Astronomers infer the presence of black holes by their dramatic effects on neighboring stars or gas clouds. A black hole can devour anything that passes too close. Its powerful gravity pulls nearby objects inward and tears them apart, accelerating and heating them up. The infalling matter emits intense X-rays (Figure 5.14).

Cygnus X-1 is an intense X-ray source over 2500 pc (8000 light-years) distant in Cygnus. Discovered in 1966, it is an eclipsing binary star (period 5.6 days) whose unseen component was the first black hole reported. The visible primary star is a blue supergiant that shows variations in spectral features from one night to the next. Presumably, when the black hole sucks in material gravitationally from its visible companion, the observed X-rays are emitted.

Recent discoveries raise new puzzles. How are high-energy jets of particles ejected from matter falling into black holes? Why are there three kinds of black holes—stellar mass, intermediate mass, and supermassive? Are black holes a source of gamma ray bursts? Look for more about these intriguing black holes in the future as scientists investigate them further.

Help astronomers discover supermassive black holes.
►radio.galaxyzoo.org◄

What do you think would happen if an unlucky spaceship passed very close to a black hole in space? _____

_____

_____

Answer: The strong gravitational pull of the black hole would pull the spaceship in, producing a destructive force that would increase as the ship fell in and that would eventually tear it apart.

SELF-TEST

This self-test is designed to show you whether or not you have mastered the material in Chapter 5. Answer each question to the best of your ability. Correct answers and review instructions are given at the end of the test.

1. Define stellar evolution. _____

   _____

2. How do astronomers check a theory of stellar evolution?_____

   _____

   _____

   _____

3. List the three main steps in the birth of a star.

   (1)_____

   (2)_____

   (3)_____

4. What is the main source of the energy that a main sequence star shines into space? _____

   _____

5. For stars of the same initial chemical composition, what property determines the length of time it takes for the stars to evolve? _____

6. Why will the Sun stop shining as a main sequence star about 5 billion years from now?_____

   _____

   _____

7. List seven stages in the life cycle of a star like our Sun in order from birth to death.

   (1) _____ (2) _____

(3) _____ (4) _____

(5) _____ (6) _____

(7) _____

8. List seven stages in the evolution of very massive stars in order from birth to death.

(1) _____ (2) _____

(3) _____ (4) _____

(5) _____ (6) _____

(7) _____

9. Why are elements that are lighter than iron, such as hydrogen, helium, carbon, and oxygen, so much more abundant in the universe than are the elements heavier than iron? _____

_____

_____

_____

_____

10. Match the eight items from the theory of stellar evolution to a real sky object.

____ (a) Birthplace of stars.   (1) Betelgeuse in Orion.
____ (b) Black hole candidate.   (2) Crab Nebula in Taurus.
____ (c) Blue giant.   (3) Crab pulsar in Taurus.
____ (d) Main sequence star.   (4) Cygnus X-1.
____ (e) Neutron star.   (5) Mira in Cetus.
____ (f) Pulsating variable star.   (6) Orion Nebula.
____ (g) Red giant.   (7) Rigel in Orion.
____ (h) Supernova remnant.   (8) Sun.

11. What is a black hole? _____

_____

_____

ANSWERS

Compare your answers to the questions on the self-test with the answers given below. If all of your answers are correct, you are ready to go on to the next chapter. If you missed any questions, review the sections indicated in parentheses following the answer. If you missed several questions, you should probably reread the entire chapter carefully.

1. The changes that take place in stars as they age—the life cycle of stars.   (Section 5.1)

2. They predict what changes in luminosity and temperature should take place in stars as they age. Then they compare these theoretical tracks of evolution on H–R diagrams with H–R diagrams for groups of real stars.   (Section 5.1)

3. (1) Gravitational contraction of a cloud of gas and dust;   (2) rise in interior temperature and pressure;   (3) nuclear fusion.   (Section 5.3)

4. Nuclear fusion reactions in the core (hydrogen is converted into helium).   (Sections 5.3, 5.5)

5. Mass.   (Section 5.4)

6. The Sun will leave the main sequence when all the available hydrogen fuel in its core is used up so that it no longer has an internal energy source.   (Sections 5.6, 5.7)

7. (1) Protostar;   (2) main sequence star;   (3) red giant;   (4) variable star;   (5) planetary nebula ejected;   (6) white dwarf;   (7) dead black dwarf.   (Sections 5.3, 5.5 through 5.13)

8. (1) Protostar;   (2) main sequence;   (3) red supergiant;   (4) variable star;   (5) Type II supernova;   (6) pulsar/neutron star;   (7) possible black hole.   (Sections 5.3, 5.5 through 5.7, 5.9, 5.14, 5.16, 5.17)

9. Hydrogen and some helium were probably the original elements in the universe. The other elements that are lighter than iron are formed inside aging stars over a period of time. Elements heavier than iron are formed only during the brief time of a supernova.   (Section 5.15)

10. (a) 6;   (b) 4;   (c) 7;   (d) 8;   (e) 3;   (f) 5;   (g) 1;   (h) 2.   (Sections 5.2, 5.5 through 5.7, 5.9, 5.14 through 5.17)

11. A superdense, gravitationally collapsed mass from which no light, matter, or signal of any kind can escape.   (Section 5.17)

6

GALAXIES

In nature's infinite book of secrecy
A little I can read.

William Shakespeare (1564–1616)
Antony and Cleopatra, Act I, ii:11

Objectives

☆ Define a galaxy.

☆ Give the observational evidence for the Milky Way Galaxy's shape, size, structure, contents, and formation, and sketch the Galaxy showing the location of the Sun.

☆ Compare and contrast open (galactic) and globular clusters.

☆ Outline the method of using H–R diagrams to determine the ages of star clusters.

☆ Describe the contents and evolution of the interstellar medium.

☆ Compare and contrast emission and absorption nebulas.

☆ Explain how maps of our Galaxy in different wavelength regions are constructed.

☆ Identify and describe the most distant object visible to the unaided eye.

☆ Compare and contrast the properties of spiral, elliptical, and irregular galaxies.

☆ Evaluate the evidence for two different models of galaxy formation and evolution.

☆ Define a cluster of galaxies and a supercluster.

☆ Give the observational evidence of large-scale structure in the universe.

☆ Compare and contrast the properties of a normal galaxy and active galaxies.

☆ Give the observed characteristics of quasars and a model that explains them.

6.1 STAR SYSTEMS

A **galaxy** is an enormous group of millions or billions of stars and gas and dust held together by the force of gravity.

Our Sun and all the visible stars in our sky belong to the **Milky Way Galaxy**. You may see a cloudy band of light across the sky on a very clear, dark night. The ancients named it the Milky Way because it looked like a trail of milk spilled in the sky by a goddess who was nursing her baby. That milky band is the combined glow of billions of stars in our huge Galaxy (Figure 6.1).

Try to locate the Milky Way overhead in summer or winter. If possible, use binoculars or a telescope to see that it is really made of many individual bright stars.

The entire Milky Way Galaxy contains over 200 billion stars. Those stars are very far apart from each other. On the average, a star's nearest neighbor star is about 5 light-years away.

What is a galaxy? _____

_____

Figure 6.1. A view toward the center of our Milky Way Galaxy in the constellation Sagittarius. Composite image combines infrared and X-ray light that see through obscuring gas and dust clouds.

Answer: An enormous collection of stars and gas and dust held together in space by the force of gravity.

6.2 MILKY WAY GALAXY

Since we are bound to the Sun, which is located inside the huge Milky Way Galaxy, we cannot photograph our own Galaxy from the outside. Instead we use photographs of distant galaxies to help us picture what our own Galaxy must look like from space (Figure 6.2).

If you could go far into space and look down on our Galaxy, you would see a brilliant spiral pinwheel about 100,000 light-years (30 kpc) across. Our Earth, traveling around the Sun, is located out in the Orion spiral arm.

From the side, you would see a thin, shiny disk with a swollen central bar over 10,000 light years (3 kpc) long that bulges in the middle. This galactic bulge is about 3000 light-years (1 kpc) thick. The average thickness of the disk is 1000 light-years (0.3 kpc). Our Sun is about 28,000 light-years (8 kpc) away from the center (Figure 6.3).

The whole Milky Way Galaxy is turning around in space. This fact is deduced from the Doppler shift of radiation from the spiral arms. Our Sun,

Figure 6.2. Spiral galaxy M51 (NGC 5194) in Canum Venaticorum resembles our Milky Way Galaxy.

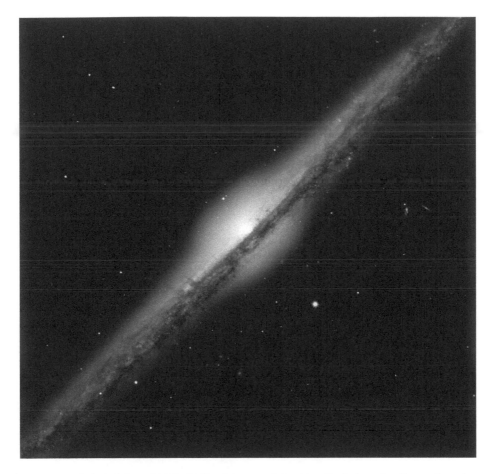

Figure 6.3. Spiral Galaxy NGC 4565, nicknamed the Needle Galaxy, in Coma Berenices. This is a spiral galaxy like our Milky Way Galaxy, viewed edge-on from Earth.

with its family of planets, is racing around the center of our Galaxy at about 250 km/sec (563,000 miles per hour). Even at that incredible speed, our solar system requires about 220 million years to complete just one revolution.

Our Galaxy appears to be hurtling through space in the direction of the constellation Hydra at a speed of over 600 km/sec (1 million miles per hour).

If you could fly across our Galaxy from one side to the other at light speed, how long would the trip take? _____

Answer: 100,000 years.

Solution: Divide the distance (100,000 light-years) by the speed (1 light-year per year).

6.3 LOCATIONS OF STARS

Our Galaxy is a barred **spiral** type. Most of the stars are concentrated in a central bar and in spiral arms that wind out from it.

While some stars travel through the Galaxy alone, many move in **star clusters**, groups of stars that stay together because of their mutual gravitational attraction. Star clusters apparently form when a gigantic cloud of gas condenses into many stars. They are important to astronomers because all the different-mass stars in a cluster are about the same age. We see evidence of the cluster origin of stars in molecular clouds that contain hundreds of thousands of solar masses.

More than a thousand **open clusters**, containing some 100 to 10,000 loosely packed stars each, have been observed. The stars move together in the disk. Open clusters are strongly concentrated in the spiral arms. Member stars are relatively young and typically hot and highly luminous (Figure 6.4).

Figure 6.4. The Pleiades (M45) open star cluster in Taurus is visible to the eye as a group of six faint stars. It has hundreds of stars and is 400 light-years away. The glow around the stars is interstellar dust, which shines by reflecting starlight.

A small fraction of the stars are in **globular clusters** in a **halo**, a spherical region around the disk. About 150 globular clusters, containing some 100,000 to 1 million tightly packed stars each, have been detected. They contain the oldest known stars (Figure 6.5).

Some globular clusters also have a small number of **blue stragglers**, atypically hot, blue, highly luminous stars. They are fusing hydrogen longer than most cluster stars because mass transfer in a binary system or collision boosted their fuel supply.

Refer to Table 6.1. List three differences between the open (galactic) clusters and the globular clusters found in our Galaxy. (1) _____

_____ ; (2) _____ ;

(3) _____

Answer: Open clusters are found in the galactic disk, are relatively young, and have a smaller number of stars. Globular clusters are found in the galactic halo, are relatively old, and have a larger number of stars.

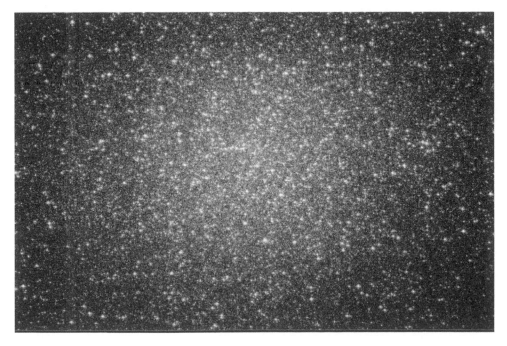

Figure 6.5. Omega Centauri (NGC 5139) in Centaurus is the biggest, brightest globular cluster. It has 10 million stars of all ages and a central black hole. At 16,000 light-years away, it looks to the eye like a fuzzy full Moon.

TABLE 6.1 Some Properties of Open and Globular Star Clusters

	Open Clusters	Globular Clusters
Location	Galactic disk	Galactic halo and galactic bulge
Diameter	Under 100 ly	Over 100 ly
Age	Relatively young	Old
Number of stars	Up to 10,000	Up to 1 million
Color of brightest stars	Blue or red	Red

6.4 THEORY CHECK

Star clusters provide the best data for verifying a theory of stellar evolution.

First, H–R diagrams predicted by theory for stars of different ages are drawn. Then H–R diagrams for observed star clusters are drawn. The theoretical and observed diagrams are compared to verify or disprove the theory.

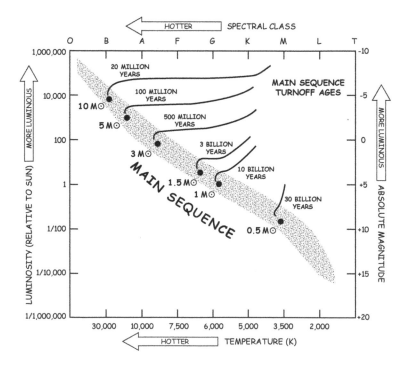

Figure 6.6. Solid lines give positions of stars in different clusters. The turnoff point away from the main sequence indicates the age of the cluster. (M⊙ = mass of Sun.)

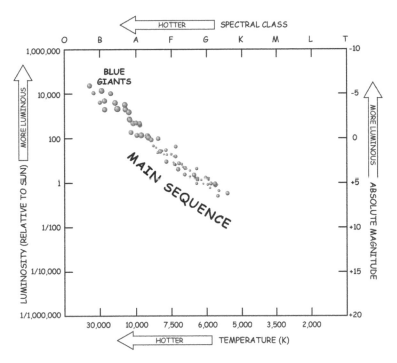

Figure 6.7. H–R diagram of Pleiades (M45) open cluster.

Figure 6.6 is a representation of predicted evolutionary tracks computed from theory. All stars start on the main sequence when they are born. The most massive stars are located at the top of the main sequence, and the least massive are at the bottom. All stars evolve away from the main sequence as they age. Massive stars evolve fastest, so the higher the turnoff point, the younger the star cluster.

Compare the H–R diagrams for the Pleiades (M45) open cluster (Figure 6.7) and the M3 globular cluster (Figure 6.8) with the theoretical evolutionary tracks (Figure 6.6). State which is a relatively (a) young cluster _____ ; (b) old cluster _____. Explain your reasoning. _____

_____

_____

Answer: (a) Pleiades cluster is relatively young. Most of its stars, even the massive short-lived ones, are still on the main sequence. (Pleiades cluster was born about 70 million years ago.) (b) M3 is relatively old. Hardly any stars appear on the upper half of the main

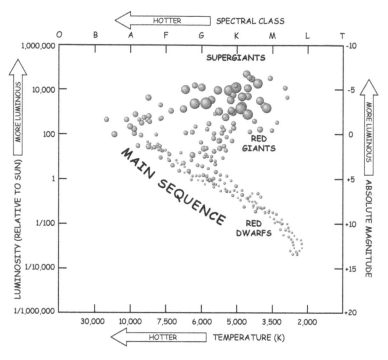

Figure 6.8. H–R diagram for globular cluster M3.

sequence, and many stars have moved to the right into the red giant region. (M3 is about 10 billion years old.)

6.5 MASS

U.S. astronomer Vera Rubin (1928–) proved that stars make up only a fraction of the mass of a **normal galaxy** such as ours. Gravitational attraction of luminous matter cannot explain observed velocities of stars and gas clouds in galaxies or galaxy clusters. Most of the matter in the universe is enigmatic **dark matter**, inferred from its gravity but unseen. Our visible Galaxy must contain a lot of dark matter and be surrounded by a huge, massive, dark matter **galactic halo** at least 300,000 light-years in diameter.

What would you expect the mass of the whole Milky Way Galaxy to be if it were concentrated in stars like our Sun? *Tip*: Use the approximate number of stars in our Galaxy from Section 6.1. _____

_____

Answer: Over 200 billion times the Sun's mass, or more than 4 x 10^{41} kg. (*Note*: The motions of stars and glowing gas indicate that the mass of the Milky Way Galaxy must be at least a trillion times the Sun's mass.)

6.6 BETWEEN THE STARS

Space is practically empty on average, but local conditions vary a lot. The **interstellar medium** is matter and radiation, pervaded by **cosmic rays** (particles moving near light speed) and magnetic fields, between the stars of a galaxy.

Interstellar matter is particularly important because it is the raw material for new stars and planets. It is about 99 percent gas (about 75 percent of the mass of the gas is hydrogen and 23 percent is helium) and 1 percent **interstellar dust**, very tiny solid particles. In our Galaxy, most of the interstellar gas and dust is concentrated in the spiral arms, and that is where the newest stars are located.

Diverse clouds of gas and dust are denser than average. They are continually enriched by material ejected by supernovas and stellar winds. Hydrogen, the most abundant gas, is in the form of neutral atoms in an **H I region**, molecules in an **H$_2$ region**, and ionized by ultraviolet rays from hot young stars in an **H II region**.

More than 150 **interstellar molecules** besides hydrogen have been detected in dense, dark, cold, giant **molecular clouds**. Water vapor and common organic molecules are the most intriguing. These are key components of all known life on Earth. Their discovery in space has raised fascinating questions about the origin of life in the universe.

Why is it important in the theory of stellar evolution to know what interstellar matter consists of in any epoch? _____

_____

Answer: Interstellar matter is the raw material for new stars and planets.

6.7 GREAT CLOUDS ⬟

Historically, **nebula**, from the Latin for "cloud," was used for all kinds of hazy patches in the sky, including many now known to be star clusters or galaxies. Today the word means a concentration of gas and dust in space.

A bright **emission nebula**, or H II region, is a cloud that glows by absorbing and then re-emitting starlight from very hot, young stars nearby. The Orion Nebula is a famous example you can observe (Figure 5.1).

Figure 6.9. The Horsehead Nebula (NGC 2024) in Orion, a famous dark nebula over 1000 light-years away, juts into bright emission nebula IC434.

A dark **(absorption) nebula**, or molecular cloud, is a relatively dense concentration of interstellar matter whose dust absorbs or scatters starlight and hides stars that are behind it from our view.

Some nebulas are given fanciful names according to their appearances. What is the "horse's head" shown in Figure 6.9 actually made of? _____

_____

Answer: Relatively dense concentrations of interstellar gas and dust.

6.8 MAPPING OUR GALAXY

We cannot look more than about a thousand light-years in most directions into our Milky Way Galaxy, even with the biggest optical telescopes, because dust clouds block our view.

Astronomers use radio, infrared, and high-energy waves, which can pass through these clouds, to image the space beyond (Figure 6.10).

The spiral structure of our Galaxy is mapped by detecting radio waves of 21-centimeter wavelength. This **21-centimeter radiation** is emitted by neutral hydrogen atoms. It is strongest from regions with the biggest concentration of neutral hydrogen atoms—the spiral arms.

Large, hot gas clouds are mapped by detecting **continuous radio emission** rather than a particular wavelength. This continuous emission comes from concentrations of excited gas in hot H II regions.

Exceptionally dense concentrations of hydrogen in dark, cool molecular clouds are harder to trace. In these regions, hydrogen atoms join together to form hydrogen molecules. Radio astronomers map the densest gas concentrations by looking at strong emission lines of carbon monoxide and other gases. Molecular hydrogen, which emits little 21-cm radiation, is observed at infrared and ultraviolet wavelengths.

Figure 6.10. A map showing the spiral structure and central bar of our Galaxy, drawn from infrared observations by the Spitzer Space Telescope.

Figure 6.11. An X-ray image and art based on X-ray and infrared images of the Galaxy's center.

New observations continually astound us. We image stellar coronas and very hot intercloud gas at ultraviolet wavelengths. Star and gas cloud motions indicate a massive dark matter halo. We detect strong radio and infrared rays and some gamma and X-rays from the galactic bulge.

The core of our Galaxy apparently contains an extraordinary, very massive, compact object ringed by very hot, chaotic gas clouds and dust called Sagittarius A*. The mysterious behavior suggests that a 4 million solar-mass black hole powers the central gas flows and luminosity. As matter falls in toward the center, it is compressed and heated to millions of degrees, producing the observed X-rays (Figure 6.11).

If, as recent observations indicate, our Galaxy has a barred spiral structure, the rate of infall would be much more rapid than in normal spirals. Violent central **starbursts**, in which great numbers of very bright and massive stars form, would occur.

What is particularly interesting about regions of relatively dense gas concentrations in our Galaxy?_____

Answer: Stars are forming in these regions.

6.9 STAR POPULATIONS

In 1944, U.S. astronomer Walter Baade (1893–1960) divided stars into two classes. Although now known to be oversimplified, this classification was useful for first explaining how age, dynamics, and element production in stars and galaxies are related.

Population I stars include the hottest and most luminous stars. These relatively young stars are located in the disk, especially in the spiral arms, embedded in the dust and gases from which they formed. They are relatively high in heavy elements (similar to the Sun, about 2 percent by mass) in addition to their hydrogen and helium.

Population II stars, like those in globular clusters, are found toward the galactic bulge and in the halo. These stars are older. They are made almost entirely of hydrogen and helium.

How does the stellar evolution theory explain the difference between Population I and Population II stars? _____

_____

_____

Answer: A star's core composition evolves, but its atmosphere's does not. Core and atmosphere usually don't mix. Population II stars formed out of primal hydrogen and helium. Population I stars formed generations later out of dust and gas enriched by elements manufactured in stars and ejected by supernovas and stellar winds.

6.10 FORMATION OF OUR GALAXY

Our Galaxy appears to have formed over 13 billion years ago, perhaps a few hundred million years after the universe began. The oldest stars are about 13 billion years old.

Theorists say our Galaxy originated in a clump of dark matter that accumulated hydrogen and helium. A turbulent cosmic gas cloud within the dark matter collapsed to star clusters and small galaxies. Mergers plus forces of gas pressure, radiation, rotation, and gravity then shaped our Galaxy to its current form.

What would this model of galaxy formation predict (a) the oldest and (b) the youngest stars in our Galaxy to be made of? Explain your answer. _____

_____

_____

_____

Answer: (a) Hydrogen and helium, the elements present as raw materials at the time our Galaxy was new. (b) Hydrogen, helium, and the other 90 naturally occurring elements. The interstellar medium is the raw material of new stars. Originally consisting of hydrogen and helium, it has been enriched by elements expelled by supernovas and stellar winds.

6.11 STRUCTURE OF THE MILKY WAY GALAXY

Refer to Figure 6.12. As a summarizing activity with regard to our Milky Way Galaxy, identify the following: (a) disk _____ ; (b) halo _____ ; (c) spiral arm _____ ; (d) galactic bulge _____ ; (e) position of Sun and Earth _____ ; (f) location of globular clusters _____

Answer: (a) 2; (b) 1; (c) 3; (d) 4; (e) 6; (f) 5.

6.12 BEYOND THE MILKY WAY GALAXY ✪

Our Galaxy was the only one recognized until 1924. Then U.S. astronomer Edwin Hubble (1889–1953) analyzed Cepheid variables and proved that some of the fuzzy "nebulas" previously observed were really distant galaxies.

The **New General Catalog (NGC)** of nonstellar astronomical objects was first published in 1888 as a list of 7840 nebular objects compiled by Danish astronomer Johann Dreyer (1852–1926). It was expanded in 1895 in a supplement, the **Index Catalog (IC)**, and again in 1908 in the **Second Index**

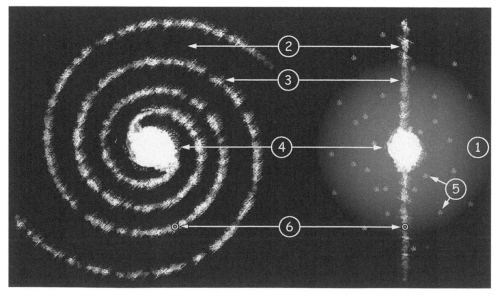

Figure 6.12. Two views of the Milky Way Galaxy.

Catalog. The **Messier Catalog (M)** of 110 nebulas, star clusters, and galaxies (Appendix 6) was originally a list of 45 hazy objects compiled in 1784 by French astronomer Charles Messier (1730–1817), who wanted to avoid mistaking them for new comets.

The universe has some 100 billion galaxies with typically over 100 billion stars each. Most bright galaxies are identified by an NGC, IC, or M number. New sightings are catalogued by their discoverers. Data and images of millions are in ▶http://nedwww.ipac.caltech.edu◀ the NASA/IPAC Extragalactic Database.

Two small, irregularly shaped galaxies, the Large Magellanic Cloud (LMC) in the constellation Tucana and the Small Magellanic Cloud (SMC) in Dorado, are famous nearby companion galaxies of our Milky Way. They and a number of faint dwarf galaxies are held around our Galaxy by the force of gravity. The LMC is at a distance of about 163,000 light-years (50 kpc) and the SMC is about 200,000 light-years (61 kpc) away, respectively (Figure 6.13).

The Magellanic Clouds are visible from the southern hemisphere. They look like small smoky clouds. Both were first noted by Portuguese explorer

Figure 6.13. The Large Magellanic Cloud in the constellation Tucana. This infrared image from the Spitzer Space Telescope is a mosaic of 300,000 tiles.

Ferdinand Magellan (c. 1480–1521) on his historic trip around the world. Both belong to a system that is inside an immense, invisible hydrogen envelope detected at 21-cm wavelength.

The Andromeda Galaxy (M31, NGC 224) is the closest galaxy that is similar to ours, with perhaps twice the mass (Figure 6.14). It is the most distant object, about 2.5 million light-years (750 kpc) away, that you can see with your unaided eye and a special sight in small telescopes. In the fall, look for a fuzzy patch of light in the Andromeda constellation (shown as O GALAXY on your star maps).

To appreciate what "close" means for galaxies, estimate how many Milky Way Galaxies you could line up next to each other between us and our neighbor, Andromeda Galaxy. _____

Answer: 25.

Solution: $\dfrac{\text{Distance to Andromeda Galaxy}}{\text{Diameter of Milky Way Galaxy}} = \dfrac{2,500,000 \text{ ly}}{100,000 \text{ ly}} = 25$

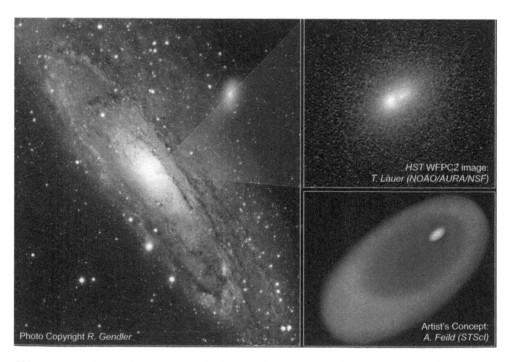

HST WFPC2 image: T. Lauer (NOAO/AURA/NSF)

Photo Copyright R. Gendler

Artist's Concept: A. Feild (STScI)

Figure 6.14. The spiral Andromeda Galaxy (M31) is even bigger than our Galaxy and contains billions of individual stars. Art shows the core of old, red stars circling a disk of young, blue stars trapped by a monster black hole's gravity.

6.13 CLASSIFICATION OF GALAXIES

Galaxies come in several different shapes and sizes. They were first classified into groups according to their structure by Edwin Hubble in 1926 (Figure 6.15).

Elliptical galaxies, designated E, are egg shaped. They range from nearly perfect spheres, E0, to the flattest, E7. Elliptical galaxies seem to contain practically all old stars. They have little visible gas and dust, but infrared and X-ray observations reveal some.

Spiral galaxies are divided into two major subcategories. **Normal spiral galaxies**, S, have a bright disk where spiral arms wind out from a bulging nucleus. They are subdivided into Sa, Sb, and Sc, according to the size of the central bulge and to how tightly wound the spiral arms are. **Barred spiral galaxies**, SB, look like normal spiral galaxies except that the spiral arms unwind from the ends of a bar-shaped concentration of material. Spiral galaxies have large amounts of gas and dust in the disk and contain young, middle-aged, and old stars.

Irregular galaxies, Irr, have no regular geometric shape. They usually contain gas and dust, mostly bright young stars and clouds of ionized gas, and some old stars.

Hubble began the systematic study of distant galaxies using the 2.5-m (100-inch) Mount Wilson telescope, the world's largest from 1918 to 1938. Today astronomers add a dizzying variety of data. **Lenticular** (lens-shaped) **galaxies**, designated S0, have bright, flat galactic disks but no spiral arms or recent star formation. **Dwarf galaxies**, small galaxies with low mass and luminosity, now seem to be the most common kind.

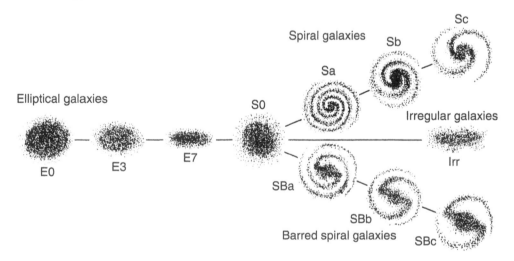

Figure 6.15. Hubble classification of galaxies according to shape and arranged in a tuning fork diagram.

Figure 6.16. Galaxies of various classifications.

Classify each of the galaxies in Figure 6.16 by its shape. (a) _____ ; (b) _____ ; (c) _____ ; (d) _____ ; (e) _____ ; (f) _____

Answer: (a) E0; (b) E4; (c) Sb; (d) Sc; (e) SBb; (f) Irr.

6.14 GALACTIC PROPERTIES

The distance to a galaxy is a key to determining the galaxy's basic properties. Distance measurements are difficult to make and are still uncertain. The uncertainties carry over to the determination of other galactic data.

A **standard candle** is an astronomical object with a known light output. We use it to determine distances to host galaxies from its known luminosity and measured brightness. Useful standard candles include Cepheid variables out to 100 million light-years (30Mpc) and standard types of galaxies out to 500 million light-years (150Mpc). Tully-Fisher relation correlates 21-cm spectral line width and luminosity of a spiral galaxy to determine absolute magnitude.

Type Ia supernovas are 4 billion times more luminous than the Sun. After using other information about their color and the rate at which they dim after their peak brightness, we get distance measurements out to 10 billion light-years precise to 10 percent.

When a galaxy's distance is determined, its luminosity and diameter can be figured from its apparent magnitude and apparent diameter.

A galaxy's mass is calculated from observed gravitational effects on its stars or gas clouds or on neighbor galaxies. Observational data indicate that most of the mass is unobserved dark matter.

Refer to Table 6.2, which summarizes rough values of the data collected so far. (Values for individual galaxies may vary a lot.) State two differences between spiral and elliptical galaxies.

_____

_____

_____

_____

Answer: Spirals contain both old and young stars; they have visible gas and dust between the stars to make new stars. Ellipticals contain old stars; they have little visible interstellar gas and dust.

TABLE 6.2 Rough Values of Galactic Data

Values	Spirals	Ellipticals	Irregulars
Mass (Milky Way Galaxy = 1)	0.005–2	0.000001–50	0.0005–0.15
Diameter (Milky Way Galaxy = 1)	0.2–1.5	0.01–5	0.05–0.25
Luminosity (Milky Way Galaxy = 1)	0.005–10	0.00005–5	0.00005–0.1
Population content of stars	Old and young	Old	Old and young
Luminous inter-stellar matter	Moderate gas and dust	Little gas and dust	Plentiful gas and dust

6.15 GALACTIC EVOLUTION

Many mysteries about galaxies still challenge astronomers.

Which formed first, galaxies or stars? **Look-back time** is the duration from when a galaxy emitted radiation to when its light reaches us. Light takes longer to travel farther. Astronomers can see ever younger galaxies in earlier eras by observing ever more distant ones.

How did galaxies form? Theory says that dark matter clumped into regions of slightly higher density in the early universe. It then collected gas that contracted rapidly to build star clusters and small galaxies that merged.

Did many galaxies go through an extremely energetic early stage in which quasars, small extraordinarily luminous objects, were their energy producing centers? Many more active galaxies are seen at great distances than nearby. How did quasars get supermassive black holes at their centers? Did the black holes cause galaxies to form or vice versa?

After birth, does a galaxy grow gradually or does it reach its final size quickly? Observations imply that for 6 billion years, galaxies grew in dynamic galaxy interactions and star birth was high.

When does a galaxy assume its shape? Initial mass, density, and angular momentum all have influence. Most likely, collisions and mergers of small galaxies over millions of years build large elliptical and spiral galaxies surrounded by a halo of stars.

How do the chemical composition, color, and luminosity of a galaxy change over billions of years? The interstellar medium evolves from initial hydrogen and helium as stars continually are born, evolve, die, and return matter enriched with heavy elements. Galaxies with a look-back time of 10 billion years (younger) are brighter and bluer than nearby (older) galaxies. Apparently hot blue stars form at a higher rate in young galaxies than in old galaxies.

Planned infrared James Webb Space Telescope (JWST) is being designed to gather key data. ▶http://www.jwst.nasa.gov◀

Refer to Figure 6.15 and Table 6.2. What observed data indicate that the different shapes of galaxies do not represent stages of evolution in the life cycle of a galaxy? _____

_____

_____

Answer: All three types of galaxies contain old stars. That fact indicates that spiral and irregular galaxies are just as old as elliptical galaxies and could not be the end stage of a galaxy's life. Nor could elliptical galaxies be the first stage of a life cycle, as proposed by

Hubble, because they do not have the dust and gas necessary for the birth of new stars seen in spiral and irregular galaxies.

6.16 GROUPINGS

Most galaxies belong to groups, called **clusters of galaxies**. These clusters contain from several to thousands of galaxies surrounded by an **intracluster** (between the galaxies) **medium**. The galaxies are held together by the force of gravity as they orbit one another at velocities of about 1000 km (600 miles) per second. **Richness** denotes the number of galaxies above a selected brightness level within a cluster. **Structure** refers to groupings of galaxies (Figure 6.17).

Our Milky Way Galaxy belongs to a typical small cluster, the **Local Group**, with more than 50 members. "Local" means that the galaxies are within a region 3 million light-years across. Three of these galaxies—our Milky Way, Andromeda (M31), and M33 in Triangulum—are spirals. The others are ellipticals (including M31's bright companions NGC 205 and M32) or irregulars (including the Magellanic Clouds). Most are dwarf galaxies, small galaxies a few thousand light-years in diameter.

Clusters divide into two classes by shape. **Regular clusters** are relatively compact, with highest density near the center. Members are mostly elliptical and SO galaxies. Many regular clusters emit radio radiation from active galaxies. About a third emit X-rays from intracluster gas at about 100 million K.

In contrast, **irregular clusters**, including our Local Group, have a looser

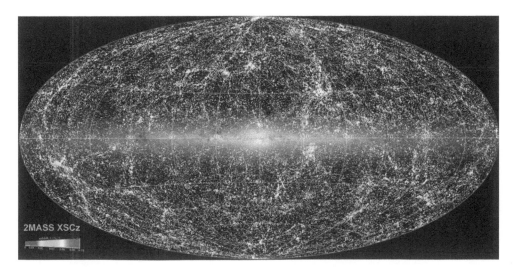

Figure 6.17. Nearly 500 million stars in our Galaxy (at center), more than 1.5 million galaxies, plus galaxy clusters and superclusters dot our local universe. Detected by the Two Micron All Sky Survey (2MASS) in infrared light.

structure with little central concentration and less very hot gas. They contain many spiral and irregular galaxies. Fewer emit radio waves or X-rays.

A **supercluster** is a cluster of clusters of galaxies. Superclusters are the largest gravitationally bound systems observed so far. They are about 100 million to 1 billion light-years across. Both the Local Group with our Milky Way Galaxy and the Virgo Cluster are members of the **Local Supercluster**.

Superclusters are located in filaments, chain-like structures, and Great Walls, thin sheets that border voids, regions where few galaxies are observed. The voids are like gigantic bubbles with clusters of galaxies along their surfaces. The observable universe consists mostly of vast voids between superclusters.

What is the largest type of structure in the universe?_____

Answer: Supercluster of galaxies.

6.17 EXTRAORDINARY ACTIVITY IN GALAXIES

Some galaxies have strange forms and unusual characteristics.

An **active galaxy** is a galaxy whose center, or **active galactic nucleus (AGN)**, emits exceptionally large amounts of energy. This energy far exceeds the total output from normal nuclear fusion reactions in stars in normal galaxies. AGNs often have great jets of hydrogen gas racing outward at very high speeds (Figure 6.18).

The colossal energy is generated around a central, powerful source of

Figure 6.18. Radiograph of Centaurus A (NGC 5128), the nearest active galaxy, 11 million light-years from Earth.

gravity that sucks in nearby matter. A black hole with a mass up to billions of times that of the Sun, is the accepted attracting force. In this model, as dust, gas, and even stars spiral in toward the black hole they accelerate and heat up. Fiery, dense, infalling matter emits the radiation. The mass of the attractor may be inferred from the speed of infall.

What is a likely explanation of the violent activity in the galaxy shown in Figure 6.18?_____

Answer: A supermassive black hole at the galaxy's center is pulling in nearby, circling stars and gas clouds.

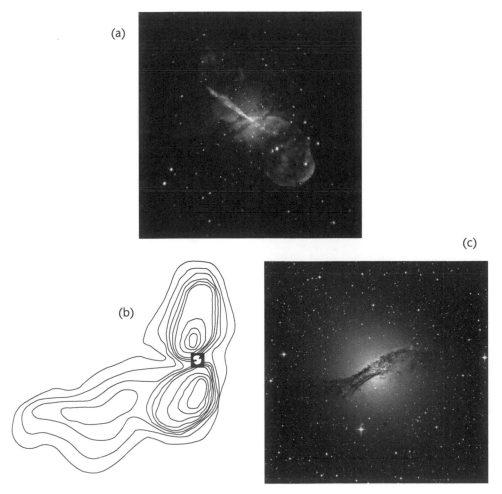

Figure 6.19. The (a) X-ray image, (b) radio contour map, and (c) visible image of Centaurus A, an elliptical galaxy with a jet of high-energy particles at its center.

6.18 RADIO GALAXIES

Active galaxies vary in looks and nature of radiation emitted.

The radiograph of a typical **radio galaxy** shows two large patches of energy at radio wavelengths on opposite sides of a visible galaxy. The radio energy usually looks like **synchrotron radiation**, or radiation produced by electrons spiraling around at nearly the speed of light in a strong magnetic field.

If the hypothetical central black hole exists, the jets of high-speed electrons are ejected by matter as it disappears into the black hole. The electrons emit the colossal radio energy while accelerating in a strong magnetic field.

Galactic close encounters could provide the prodigious matter that the hypothetical black hole devours.

When two galaxies collide, they apparently pass through each other. New observations of the Sagittarius dwarf galaxy, which orbits our Milky Way, show it plunges into the central regions of the Galaxy. The clouds of gas and

Figure 6.20. Artist's concept. Titanic collision and merger of our Milky Way with neighboring Andromeda Galaxy 4 billion years from now.

dust in **colliding galaxies** would be unusually dense and could trigger a burst of star formation or fuel the hypothetical black hole (Figure 6.20).

Galactic cannibalism probably occurs when a very large galaxy passes too close to a much smaller one and "eats" it. A massive galaxy could tidally strip away and take in gas, dust, and stars from a smaller disk. The smaller nucleus, falling to the center of the massive galaxy, would fuel the larger's energy output for millions of years.

What might happen to life on Earth when our Galaxy collides with Andromeda billions of years from now? Explain. _____

_____

_____

Answer: Probably nothing. Although our Milky Way and Andromeda are approaching each other at 110 km/sec (246,000 miles per hour), their stars and planets are too far apart inside to collide. Only the disks, central black holes, and interstellar clouds will merge.

6.19 INFRARED GALAXIES

Infrared galaxies are active galaxies that shine brightest at infrared wavelengths.

About 1 percent of all spiral galaxies are **Seyfert galaxies**, first described by U.S. astronomer Carl K. Seyfert (1911–1960) (Figure 6.21).

A **Seyfert nucleus**, typically only 10 light-years across, shines many times more brilliantly than a normal spiral galaxy. Its spectrum has broad emission lines that indicate turbulent motions of very hot gas at velocities of thousands of kilometers per second.

Heated dust enveloping the nucleus probably absorbs high-energy radiation emitted from the energized core and re-emits it at longer infrared wavelengths.

A **starburst galaxy**, one in which stars are forming at a rate hundreds of times greater than in our Galaxy, often shines strongest at infrared wavelengths. Some have two different galactic centers, which indicates they result from a collision of two gas-rich spiral galaxies.

Why might a starburst galaxy emit most of its energy at infrared wavelengths? _____

_____

_____

_____

Figure 6.21. The lumpy, thick ring around the nucleus of Seyfert galaxy NGC 7742 is an area of active star birth. Tightly wound spiral arms are faintly visible.

Answer: The galaxy may result from two gas-rich galaxies' collision and so have an unusually high rate of star formation. Nearby dust absorbs the hot young stars' ultraviolet rays and re-emits infrared rays.

6.20 MYSTERIOUS QUASARS

The first quasars observed look like faint stars in photographs but are strong radio sources with nonstellar spectra, hence the name quasi-stellar radio source and its contraction to quasar (Figure 6.22).

Most of the thousands of quasars emit extraordinary power across a broad range of wavelengths, from radio to gamma rays. Although most are not strong radio sources, the original name stuck. A quasar is the most luminous active galactic nucleus.

Quasars are extremely compact, typically about 1 light-day across (not much bigger than our solar system), but they shine brighter than a thousand

(a)   (b)

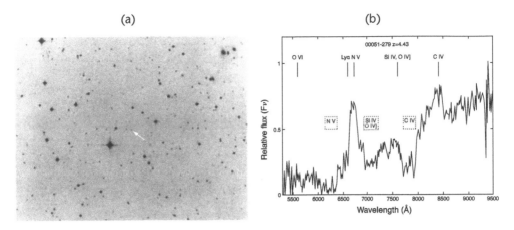

Figure 6.22. Extremely distant Quasar Q0051-279 (a) is practically indistinguishable from stars in ordinary photographs. (b) Its spectrum has light so far shifted to the red as to place this quasar near the assumed beginning of the universe.

normal galaxies. Most quasars vary irregularly in their light output. A **blazar**, a quasar with jets pointed at Earth, varies most.

Quasars have the highest redshifts observed. This high redshift, symbol **z**, indicates that quasars are racing away from us at speeds of over 90 percent the speed of light. Ultraviolet light emitted by quasars with large redshift is received as red light on Earth. This redshift implies that quasars are very distant objects. They were shining when the universe was young.

Observations of their environs confirm that quasars are extremely far away. CCD detectors show spectrums of quasars, faint haze around the quasars, and remote galaxies that all look identical. Brilliant, compact quasars must be at the centers of very distant galaxies.

Twin or multiple images of the same quasar, formed by a **gravitational lens**, a concentration of mass that bends the path of light, provide further evidence that quasars are located at cosmological distances. According to Einstein's **general relativity** theory, a theory of gravitation, starlight passing near a massive body is deflected. Galaxies, intracluster gas, or dark matter can act as a gravitational lens and let us obtain images and spectrums of outlying objects too faint to examine otherwise (Figure 6.23).

Different hypotheses were proposed and abandoned to explain the stupendous energy output of these small cosmic powerhouses. Einstein's theory of relativity was used to attribute the quasars' extraordinary redshift to an enormous gravitational force **(a gravitational redshift)**, which would have meant the quasars were closer and not so powerful after all. Collisions of material particles with **antimatter**, exotic opposite of matter on Earth, were invoked.

The evidence shows that a quasar is powered by a supermassive black

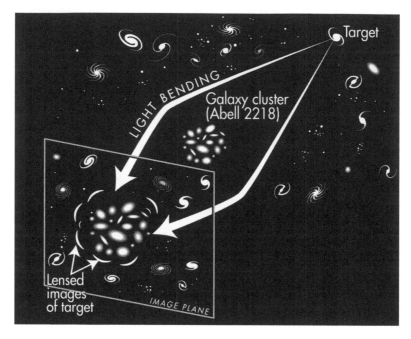

Figure 6.23. Gravitational lens. A galaxy cluster acts like a giant magnifying glass. It bends and focuses light rays (bent arrows) from a distant galaxy (target) into several magnified and distorted images of the distant galaxy.

hole. Since quasar activity was much more common in the early universe than it is today, a quasar must be a development phase in young galaxies.

Because they are so distant, quasars are useful probes of very distant hydrogen clouds. When quasar light passes through intergalactic clouds, some is absorbed. Absorption lines in the quasar's spectrum hold clues to the birthplace of galaxies in the early universe.

Astronomers catalog and analyze the sizes, shapes, brightnesses, colors, redshifts, and distributions of numerous galaxies and quasars to better understand the universe.

Help astronomers understand galaxies at ▶www.galaxyzoo.org◀

What is a quasar? _____

_____

Answer: The very luminous active center of a distant galaxy.

SELF-TEST

This self-test is designed to show you whether or not you have mastered the material in Chapter 6. Answer each question to the best of your ability. Correct answers and review instructions are given at the end of the test.

1. Define a galaxy. _____
   _____

2. Arrange the following in order of increasing size: star, planet, galaxy, cluster of galaxies, open cluster, supercluster, solar system. _____
   _____

3. Sketch an edge-on view of the Milky Way Galaxy, and label the (a) size of the diameter; (b) disk; (c) galactic bulge; (d) spiral arm; (e) halo; (f) position of Sun and Earth; (g) location of globular clusters._____

4. Which of the following have been identified in the interstellar medium: hydrogen gas, radiation, bacteria, tiny solid dust particles, viruses, water vapor, spirits, gases of elements heavier than hydrogen, organic molecules, algae?_____
   _____

5. Why is it important in the theory of stellar evolution to know what interstellar matter consists of in any epoch?_____
   _____

6. Refer to Figure 6.9. Is the space behind the "horse's head" really empty of stars? Explain. _____
   _____
   _____
   _____
   _____

7. Why is the 21-cm radio radiation emitted by hydrogen atoms more useful than visible light in mapping the structure of our Milky Way Galaxy?

_____

_____

8. Refer to the H–R diagrams of star clusters 1 and 2 in Figures 6.24a and 6.24b.

(a) Which cluster is older? _____

(b) Which cluster would have population I stars?_____

(c) Which cluster would have stars with a relatively high abundance of heavy elements?_____

(d) Which is a globular cluster? _____

Figure 6.24a. H–R diagram of "Cluster 1."

Multi-wavelength images of the Sun from the TRACE and SOHO spacecraft.

Montage of the planets and smaller bodies in the
solar system.

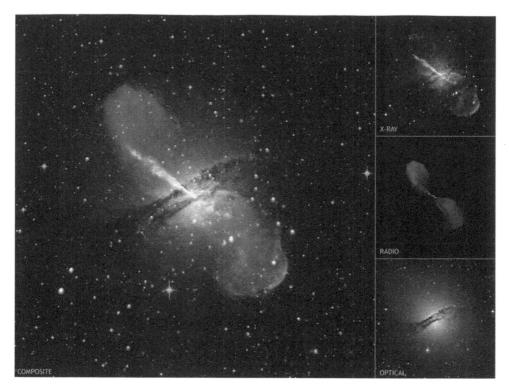

X-RAY

RADIO

OPTICAL

COMPOSITE

False color composite of galaxy Centaurus A from X-ray, optical, and radio data shows the effects of an active supermassive black hole.

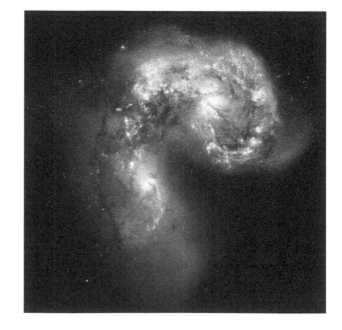

Antennae Galaxies. An interacting pair whose streamers of stars, about a hundred thousand light-years long, were created when they collided.

Star-forming region NGC 3603, about 20,000 light-years away in the Carina spiral arm of our Galaxy.

Expanding halo of light in interstellar dust around red giant star V838 Monocerotis (V838 Mon), about 20,000 light-years away in the constellation Monoceros.

Detailed, all-sky picture of the infant universe from Planck data reveals 13.8 billion-year-old temperature fluctuations (shown as color differences).

Whirlpool Galaxy (M51). False color composite from multi-wavelength data has purple (X-ray) black holes and neutron stars in binary star systems; green (optical) and red (infrared) lanes of stars, gas, and dust; and blue (ultraviolet) hot, young stars.

Edge-on view of the Sombrero galaxy (M104). False color shows blue (X-ray) hot gas, objects within, and quasars, and green (optical) and red (infrared) rim of dust around a central bulge of stars.

(e) Which has many bright blue stars? _____

(f) Which may contain up to 10 million stars? _____

9. (a) What is the most distant object visible to the unaided eye? _____

_____

_____

(b) How long does it take light emitted from that object to reach your eyes?

_____

_____

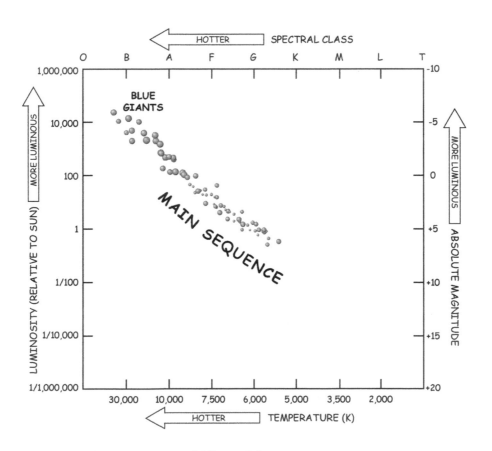

Figure 6.24b. H–R diagram of "Cluster 2."

10. List the main shapes of galaxies in the Hubble classification scheme, and explain why they cannot represent successive stages of galaxies' evolution.

_____

_____

11. What is the most popular explanation for the colossal energy output of active galaxies? _____

_____

12. Match the following descriptions with the correct object:

_____ (a) Shows the largest redshift known.

_____ (b) The clouds of gas and dust here are much more dense.

_____ (c) Radiograph shows two large patches emitting radio waves on opposite sides of a visible galaxy located between them.

_____ (d) Has a relatively small brilliant nucleus with broad emission lines in its spectrum.

_____ (e) Its luminosity can be explained as the composite of a collection of many individual stars.

(1) Colliding galaxies.
(2) Seyfert galaxy.
(3) Normal galaxy.
(4) Quasar.
(5) Radio galaxy.

ANSWERS

Compare your answers to the questions on the self-test with the answers given below. If all of your answers are correct, you are ready to go on to the next chapter. If you missed any questions, review the sections indicated in parentheses following the answer. If you missed several questions, you should probably reread the entire chapter carefully.

1. An enormous collection of stars and gas and dust held together in space by the force of gravity.   (Section 6.1)

2. Planet, star, solar system, open cluster, galaxy, cluster of galaxies, supercluster.   (Sections 6.1 through 6.3, 6.16)

3. Figure 6.25.   (Sections 6.2, 6.3, 6.11)

Figure 6.25. Edge-on view of the Milky Way.

4. Hydrogen gas, radiation, tiny solid dust particles, water vapor, gases of elements heavier than hydrogen, organic molecules.   (Section 6.6)

5. Interstellar matter is the raw material for new stars and planets.   (Sections 6.6, 6.8)

6. No. The "horse's head" is a dark nebula. It is a relatively dense concentration of interstellar matter whose dust absorbs or scatters starlight and hides stars that are behind it from our view.   (Section 6.7)

7. Radio waves pass through the interstellar dust in the disk of the Milky Way Galaxy much more effectively than visible light waves.   (Section 6.8)

8. (a) 1;   (b) 2;   (c) 2;   (d) 1;   (e) 2;   (f) 1.   (Sections 6.3, 6.4, 6.9)

9. (a) Andromeda Galaxy;   (b) around 2.5 million years.   (Section 6.12)

10. Elliptical, spiral, irregular. All contain old stars, so all must be equally old.   (Sections 6.13 through 6.15)

11. A supermassive black hole at the galaxy's center. (Section 6.17)

12. (a) 4;   (b) 1;   (c) 5;   (d) 2;   (e) 3.   (Sections 6.5, 6.17 through 6.20)

7

THE UNIVERSE

*In the beginning God created the heaven and the Earth.
Now the Earth was unformed and void, and darkness was
upon the face of the deep; and the spirit of God hovered
over the face of the waters. And God said: "Let there be
light." And there was light. And God saw the light, that it
was good.*

Genesis 1:1–4

Objectives

☆ Define cosmology.
☆ Describe the basic assumptions and limitations of cosmology.
☆ Specify the evidence that the universe is expanding.
☆ State the Hubble law.
☆ Explain the significance of the Hubble constant.
☆ Describe the past and present of the universe according to the Big Bang theory.
☆ Compare and contrast the future of the universe according to the open, flat, and closed models of the universe.
☆ List important observations in support of the Big Bang theory.
☆ Describe methods for choosing among the open, flat, and closed models of the universe.
☆ Outline a problem of the standard Big Bang model and its resolution by the inflationary universe model.
☆ Describe astronomical methods of estimating the age and size of the universe.

7.1 ETERNAL QUESTIONS

People have always wondered about how the world began and if it will end. Ancient myths, philosophy, and theology all provide models. **Cosmology** is the study of the origin, present structure, evolution, and destiny of the universe.

Astronomers construct **cosmological models**, mathematical descriptions that try to explain how the universe began, how it is changing as time goes by, and what will happen to it in the future. These models must be consistent with the observational data we have on stars and galaxies.

Two basic types of models, **evolutionary** and **steady state**, have been tested in the last 50 years. Results substantiate the evolutionary type.

Cosmological models differ from religious explanations of the universe in a fundamental way. Can you state the difference? _____

_____

Answer: Cosmological models do not give a supernatural cause or meaning to physical events but try to explain these events using only the laws of nature and mathematics.

7.2 THE EXPANDING UNIVERSE

The basic observation that must be accounted for by any cosmological model is that light from distant galaxies is shifted in wavelength toward the red end (longer wavelengths) of the spectrum. This phenomenon is called the **cosmological redshift**.

Modern theory says this redshift, symbol **z**, results from an expansion of space. As distances between galaxy clusters increase, traveling light waves are stretched. Greater redshifts correspond to more distant galaxies and earlier eras. The most distant, youngest galaxies we observe have the greatest redshifts. They are moving away fastest of all (Figure 7.1).

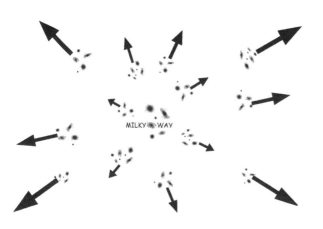

MILKY WAY

Figure 7.1. Galaxy clusters receding, as they appear from our Milky Way Galaxy (arrows indicate velocities).

When we look into space we see galaxy clusters receding from us. What does this observation imply about the universe? _____

_____

Answer: The universe must be expanding. (Gravity binds galaxies within the clusters.)

7.3 REDSHIFTS

Examine Figure 7.2, which shows redshifts and corresponding calculated velocities for five galaxies that are at different distances from us.

Two spectral lines are shown on the right for each galaxy pictured on

Figure 7.2. The Hubble law. Redshifts and corresponding velocities of five galaxies.

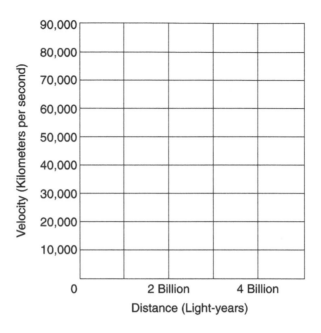

Figure 7.3a. Grid for velocity versus distance diagram.

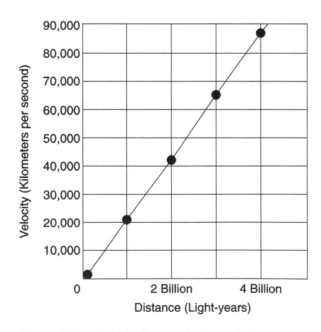

Figure 7.3b. Hubble diagram for five galaxies.

the left. More distant galaxies look smaller. A pair of the darkest absorption lines, H and K of ionized calcium, are distinctive and easy to spot. An arrow is marked at the top left of each strip to point out that these two lines are shifted toward the red by increasing amounts for more distant galaxies.

Use Figure 7.3a. Plot a rough graph where each point represents the velocity of recession and distance of one of these galaxies. What do you observe when you draw a smooth curve through the five points? Explain._____

_____

_____

Answer: The points all lie near a straight line (Figure 7.3b). That means there is a linear relationship between velocity of recession and distance from us for these galaxies. (We don't yet know about the most distant galaxies.)

7.4 VELOCITY-DISTANCE RELATION

U.S. astronomer Edwin Hubble, who spent most of his life studying galaxies, examined the relationship between the velocity of recession and the distance away for many galaxies. He discovered that the linear relationship you just found is true in general: the farther away a galaxy is, the faster it is receding.

The **Hubble law** (1929) says that a galaxy's velocity of recession, v, is directly proportional to its distance from us, d. The formula is:

$$v = H_0 d \text{ where } H_0 \text{ is called the Hubble constant.}$$

The Hubble constant is very important. It gives the rate at which the galaxies are receding, or the rate at which the universe is now expanding. It is also used in the Hubble law to figure the distance to galaxies from their measured redshifts.

To determine H accurately is difficult because of uncertainties in the extragalactic distance scale. The best current value is $H_0 = $ 69 km/sec/Mpc (21 km/sec/MLY) $\pm$ 1% from independent measurements by space telescopes and cosmology satellites. Expect more precision as research continues.

Some primitive galaxies have the largest redshifts ever observed. If this phenomenon is due to the expansion of the universe and they are moving away faster than any nearby galaxy, what can you say about their distance? Explain.___

_____

_____

Answer: These primitive galaxies are the most distant objects we observe. The Hubble law says the most distant objects are those that are moving away the fastest.

7.5 PREMISE

The basic assumption we make in attempting to understand the universe is called the **cosmological principle**.

The cosmological principle states that on a sufficiently large scale the universe is **homogeneous** and **isotropic**. At any given time, the distribution of matter is the same everywhere in space, and the universe looks the same in all directions.

Our neighborhood in space is not particularly special. The laws of physics are universal. Any observer anywhere in the universe would see at any given time just about the same things we do on a large scale.

The cosmological principle is important because it lets us assume that the small portion of space we can see is truly representative of all the rest of the universe that we cannot see. It allows us to formulate a theory that explains the entire universe, including those parts we cannot observe.

Our observations show that other galaxy clusters are racing away from us. Does that mean that our Milky Way Galaxy is the center of the entire universe? Explain.

_____

_____

Answer: No. The cosmological principle says that if you went to any other galaxy and looked very far out into space, you would still see approximately equal numbers of galaxy clusters located in all directions in space, racing away from you.

You can do a simple activity to illustrate this principle (Figure 7.4). Obtain a balloon and let its skin represent three-dimensional space. Put stickers randomly on the balloon to depict galaxy clusters and sketch a wavy line for a photon. Label one sticker LG to represent our Local Group. Blow up the balloon. Observe how all the stickers (galaxy clusters) are pulled increasingly farther apart (recede from one another) and the wavy line (wavelength of radiation) stretches (cosmological redshift) as the balloon's skin expands.

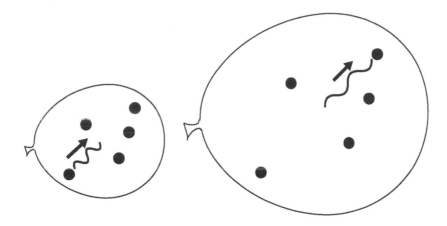

Figure 7.4. An inflating balloon provides a conceptual model of the expanding universe.

7.6 STANDARD BIG BANG THEORY

The **Big Bang theory** says that some 13.8 billion years ago our universe expanded rapidly from an infinitely hot, dense speck and it has been evolving ever since. The Big Bang was the beginning of all of time and space we can know .

All of the matter and radiation of our present universe were packed together at the start. At 10^{-43} seconds after the Big Bang, the temperature was 10^{32} K. The early universe was opaque, made of a nearly featureless hot, charged gas that emitted and trapped high energy photons of light.

Expansion cooled the matter and photons of that early inferno. In a few seconds, protons (hydrogen nuclei), neutrons, electrons, positrons, and neutrinos formed. Within minutes, deuterium (heavy hydrogen), helium, and a few lithium nuclei formed.

After 380,000 years, the expanding universe cooled enough for electrons and nuclei to combine into neutral atoms. Matter and radiation were decoupled. Photons sped across the universe.

Several million years later, stars and galaxies began to form. The universe has continued to expand in space-time, the galaxies have continued moving apart, and the radiation has continued to cool ever since.

Today we observe the universe still expanding. Stars are still forming inside galaxies, using the original hydrogen from the Big Bang. The observed material of the universe is approximately 74 percent hydrogen and 24 percent helium, with traces of other light elements, such as deuterium and lithium, as predicted.

Predictions vary about the future, when the original hydrogen will finally be used up in stars and they will all stop shining. The fate and shape of the

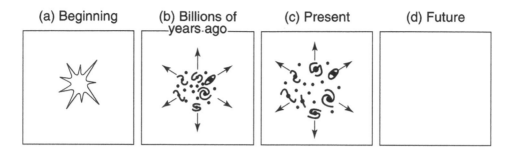

Figure 7.5. Stages of an open and a flat universe (Big Bang theory).

universe will be determined by the competition between the outward expansion and the inward pull of gravity.

If these are equal, then the universe is flat like a card and infinite. If gravity is weaker, space is open (endless) and negatively curved like a saddle. Either way, the universe will continue to expand indefinitely.

Refer to Figure 7.5. Identify and briefly describe the stages of an open and a flat universe according to the Big Bang theory if gravity acts alone.

(a) _____ ; (b) _____ ;

(c) _____ ; (d) _____

_____ .

Answer: (a) Big Bang explosion took place. (b) Galaxies formed. (c) Galaxies are still receding; the universe is expanding. (d) Original hydrogen will be used up; the resulting cold, black universe will continue expanding and decelerating indefinitely.

7.7 BIG CRUNCH

If gravity is stronger, gravity wins. The geometry of space is closed and positively curved like a sphere. The universe will not expand forever.

If the universe is closed, we happen today to be in the observed expanding phase. In the future the expansion will slow down. Gravity will halt the expansion and force a collapse. Galaxies will fall back inward toward one another until all matter is once again crunched into an extremely hot, dense state—the Big Crunch.

The **oscillating universe** model variation says that after the Big Crunch, another Big Bang will occur. Then a new expanding universe will be born out of the same matter. The universe oscillates forever.

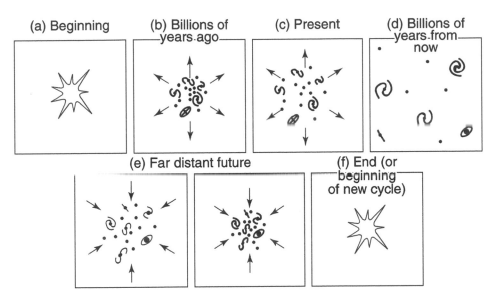

Figure 7.6. Stages of the closed (or oscillating) universe (Big Bang theory).

Refer to Figure 7.6. Briefly describe the stages of a closed universe according to the Big Bang theory if gravity acts alone. (a) _____ ;

(b) _____ ;

(c) _____ ;

(d) _____ ;

(e) _____ ;

(f) _____ .

Answer: (a) Big Bang occurred. (b) Galaxies formed and continued to recede. (c) We live in an expanding universe; galaxy clusters are racing away from one another today. (d) Galaxies will stop. (e) The universe will contract; galaxies will fall back inward. (f) Matter will be crunched together again.

7.8 STEADY STATE THEORY

The **Steady State theory**, proposed by UK astronomer Sir Fred Hoyle (1915–2001) in 1948, was a rival of the Big Bang theory a few decades ago.

This theory said that the universe does not evolve or change in time. There was no beginning in the past and there will be no end in the future. Past, present, future—the universe is the same forever.

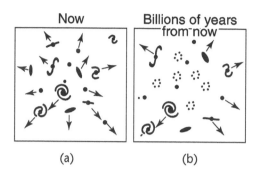

Figure 7.7. Stages of the universe (Steady State theory).

The Steady State theory assumes the **perfect cosmological principle**, which says that the universe is the same everywhere on the large scale, at all times. It maintains the same average density of matter forever.

In order to explain the observation that the universe is expanding, the Steady State model says that new hydrogen is created continuously in empty space at a rate just sufficient to replace matter carried away by receding galaxies. The theory does not explain where the new hydrogen comes from.

Astronomers reject the Steady State theory because it contradicts observations. They accept only models that observations support and experiments other scientists confirm.

To its proponents, the Steady State theory had philosophical appeal. It defined a universe that always existed in the past and will always exist in the future. The hypothesis that nuclear reactions in red giant stars produced chemical elements from carbon to iron and that heavier elements were made in supernovas, advanced to explain how they got here without a Big Bang, survives intact in evolutionary models.

Refer to Figure 7.7. Briefly describe the stages of the universe according to the Steady State theory. (a) _____

_____

(b) _____

_____

Answer: (a) Galaxies are receding, the universe is expanding, new matter is being created, new galaxies are being formed. (b) The same pattern will occur. The universe maintains the same average density forever.

7.9 OBSERVATIONAL TESTS

Astronomers test a cosmological model by seeing whether it agrees with all the observational data we have about the universe.

The most direct way to check how the universe is evolving is to compare the way it looks today with the way it looked billions of years ago. Since we cannot actually make observations over billions of years as the universe ages, astronomers instead look at galaxies that are at different distances away from us.

Although the idea—to look back in time you study distant galaxies—is simple, it is very hard to carry out in practice. We can measure redshifts readily, but technology is not sufficiently developed for precise distance measurements.

Consequently, all data that might be used to check cosmological models have uncertainties. Distant galaxies (longer look-back times) differ from nearby galaxies, which confirms that our universe evolves. New data imply that our universe is flat. Tests are ongoing.

How can astronomers find out what the universe was like (a) 2 million years ago? (b) 3 billion years ago? Explain. _____

_____

_____

_____

Answer: They examine radiation emitted by distant galaxies such as (a) Andromeda, which is about 2.5 million light-years away from Earth, and (b) Hydra, which is about 3 billion light-years away. It takes light one year to travel a distance of 1 light-year. The light we now receive left Andromeda 2.5 million years ago or Hydra 3 billion years ago; it tells us now what the universe was like then.

7.10 INCONSTANT HUBBLE CONSTANT

Cosmologists compare the value of the Hubble constant now and billions of years ago to verify their predictions.

Evidently the Hubble constant is not constant over time. The expansion of the universe decelerated in the past and is accelerating now.

Theory predicts how far distant galaxies would be from Earth if the expansion has been slowing down or constant.

Observations of Type 1a supernovas in distant galaxies show that they are fainter than expected from redshift data. This indicates that the expansion of the universe began accelerating a few billion years ago.

The supernova data mean that there is strong negative pressure that opposes and overcomes gravity. Our universe is full of **dark energy**, an as yet unknown source of gravitational repulsion.

Hubble constant data are not yet exact because no one can measure the distances to galaxies precisely.

Why is it so important to measure the value of the Hubble constant very accurately? _____

_____

Answer: An accurate value of the Hubble constant would be strong evidence in favor of one of the cosmological models described. The Hubble constant is used to figure the age and size of the universe.

7.11 MATTER AND ENERGY

General relativity says that the average density of our universe determines its shape.

The **critical density** is the minimum average density of matter and energy required to make the universe flat. The calculated value, which depends on the value of the Hubble constant (uncertain), is roughly 10^{-30} grams per cubic centimeter or a few hydrogen atoms per cubic meter.

The abundance of the lightest elements in space today sets a limit to the maximum possible amount of ordinary matter in our universe. Essentially all existing hydrogen, helium, and lithium is presumed to have been created at the Big Bang, so it is tightly linked to the initial density of matter. Observations indicate that the composition of the universe is only about 4.9 percent ordinary matter, 26.8 percent dark matter of the critical value. The resulting gravitational force is less than necessary to eventually halt the observed expansion.

Omega, Ω, the ratio of the average density of all matter observed by its radiation or gravitational effect to the critical value, is now less than 1. If the universe is flat as theorized, the rest must be made of 68.3 percent dark energy.

Dark matter and dark energy as yet undetected are necessary for a critical average density of matter. Massive neutrinos, **MACHOs**—massive compact halo objects—or **WIMPs**—weakly interacting massive particles—may exist. A **cosmological constant**, Λ, dubbed dark energy and equivalent to a cosmic repulsion, was introduced in general relativity by Einstein and may drive an ever-accelerating expansion.

What would be the cosmological significance of new discoveries of so-far-unobserved dark matter and dark energy in the universe? _____

_____

_____

Answer: The density of matter in the universe would be closer to the critical value, evidencing that our universe is nearly flat.

7.12 UNIVERSAL RADIATION

The Big Bang theory predicts that the universe should be filled with cosmic microwave background radiation, CMB, a vestige of its superhot origin.

The primeval fireball would have created strong shortwave radiation like exploding gigantic atom bombs. After 380,000 years, that radiation (equivalent to a temperature of 3000 K) would spread out, cool, and fill the expanding universe uniformly. By now, stretched by the expansion, it would strike Earth as microwaves, corresponding to a temperature of 2.73 K.

In 1965 U.S. physicists Arno Penzias and Robert Wilson detected microwaves coming equally from all directions in the sky, day and night, all year. Since 1989, data from cosmological satellites have matched this nearly uniform radiation to that of a blackbody at exactly 2.73 K.

Astronomers have detected and analyzed the remnant of the Big Bang radiation. Its random irregularities are key to the origin of large-scale structure in the universe.

What did the discovery of the cosmic microwave background mean for the Steady State theory? _____

_____

Answer: It invalidated the Steady State model, since the model cannot explain the existence of this radiation.

7.13 BIG BANG MODEL SUCCESSES

Summarize three observations that support the Big Bang model.

_____

_____

_____

_____

Answer: Your answer should include the following observations: (1) redshifts of distant galaxies (universe is expanding); (2) cosmic microwave background (remnant of heat from Big Bang); (3) abundances of hydrogen, helium, and lithium (lightest elements formed in the first few minutes).

7.14 BIG BANG QUESTIONS

The original Big Bang model fails to explain how an explosive beginning resulted in both the near homogeneity of the cosmic microwave background and the large-scale structure of the observable universe.

If the initial distribution of energy and mass was smooth and the expansion gradual, gravity alone could not clump ordinary matter into the observed large clusters and superclusters of galaxies in the calculated lifetime of the universe. That lifetime is also too short for widely separated regions of space that are observed to be similar today to ever have had a link, because they are too far apart to communicate even at light-speed.

In 1981, U.S. physicist Alan Guth proposed adding inflation, a brief phase of incredibly rapid expansion shortly after the Big Bang, to explain why the earliest density of the universe must have been extraordinarily close to the critical density. The expansion accelerates when exotic mass-energy repels gravity.

Tiny initial anisotropies and inhomogeneities, or quantum fluctuations, in the energy density were in close contact initially. Inflation spread them into the whole universe, providing the seeds whose gravitational pull grew the observed galaxies, clusters of galaxies, and superclusters in space today.

Cosmological satellites, ESA/NASA Planck (2010–2013) most recently, measured and mapped slight temperature variations in the average CMB temperature of 2.73 K. They evidence the tiny initial fluctuations.

Inflationary cosmology calls for a homogenous, isotropic, flat universe. Proponents hypothesize that an enormous amount of dark matter and dark energy in the universe is exotic and escaping detection.

Few astronomers want to abandon Big Bang theory entirely. What are two major concerns today? (1) _____

_____

(2) _____

_____

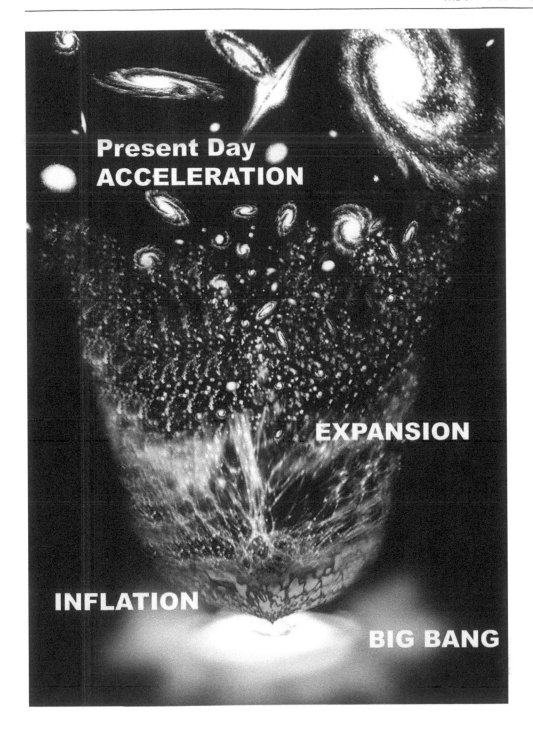

Figure 7.8. History of the universe from the Big Bang to the present.

Answer: (1) The value of the expansion rate of the universe (Hubble constant). (2) Unambiguous detection of dark matter and dark energy, in a familiar or exotic form.

7.15 AGE

Estimates of the age of the universe have tended to grow from a biblical few thousand years to millions and then billions of years.

Today astronomers (1) measure the rate of expansion and extrapolate back to the Big Bang and (2) date the oldest stars, which place a lower limit to the age of the universe.

The **Hubble time**, the age of the universe since the time of the Big Bang, is equal to 1/H. The calculated age depends greatly on the value of the (still imprecise) Hubble constant, with correction for the slowing down of the universe in the past and accelerating now. In 2013 Planck measurements put the time of the Big Bang at 13.8 billion years ago.

Hubble Space Telescope observations of white dwarf stars in the oldest globular clusters yield a similar age. White dwarfs cool down at a predictable rate—the older the dwarf, the cooler it is. Some measured 12 to 13 billion years old. Since the first stars formed about 400 million years after the Big Bang, this method provides confirmation.

List two methods of estimating the age of the universe. (1)_____

_____

(2) _____

Answer: (1) Measuring the Hubble constant and Hubble time of 1/H; (2) estimating based on the age of the oldest stars observed.

7.16 SHAPE AND SIZE

The cosmic microwave background holds details about the history, shape, content, and ultimate fate of the universe.

Cosmic mass and energy curve space, according to general relativity theory. Planck's measurements of cosmic microwave background fluctuations indicate that the shape of the observable universe is flat, like a piece of paper (Figure 7. 9).

The size of the observable universe is the distance that light has had time to travel since the Big Bang. The **Hubble radius**, equal to the speed of

Figure 7.9. The angular sizes of the fluctuations in the cosmic microwave background depend on whether the curvature of space is (a) negative (open), (b) zero (flat), or (c) positive (closed). Today measurements indicate a nearly flat universe.

light divided by the Hubble constant, c/H_0, is about 14 billion light-years. It increases as the universe ages.

What are the best figures we have for the observable universe today (latest observations)?

(a) Rate of recession of distant galaxies (Hubble constant) _____

_____

(b) Approximate radius _____

(c) Age in approximately present form _____

Answer: (a) The rate of recession (Hubble constant) is 69 ± 1% km/sec/Mpc (21 km per second per million light-years). (b) About 14 billion light-years. (c) 13.8 billion years old.

SELF-TEST

This self-test is designed to show you whether or not you have mastered the material in Chapter 7. Answer each question to the best of your ability. Correct answers and review instructions are given at the end of the test.

1. Define cosmology. _____

   _____

2. How do cosmological models differ from religious explanations of the universe? _____

   _____

   _____

3. Describe the evidence that the universe is expanding. _____

   _____

   _____

   _____

   _____

4. State the Hubble law. _____

   _____

   _____

   _____

   _____

5. Why is the Hubble constant so important in cosmology?_____

   _____

   _____

   _____

6. State the basic assumption of all cosmological models. _____

   _____

   _____

7. Match one or more of the three main cosmological models (Big Bang theory) with each statement:

    ____ (a) 13.8 billion years ago the universe exploded into being from an extremely dense hot state.

    ____ (b) An enormous amount of matter and energy in the universe is dark, exotic, and escaping detection.

    ____ (c) Galaxies are moving away from each other with speeds that increase with distance.

    ____ (d) In the future, the universe will expand indefinitely.

    ____ (e) In the future, the universe will stop expanding and then contract.

(1) Open.

(2) Closed.

(3) Flat.

8. List two basic observations that can help decide the ultimate fate of the universe. (1) _____ ;
(2) _____

9. What is the cosmological significance of cosmic microwave background?

_____

_____

_____

10. State the main contribution of an inflationary universe model to Big Bang theory. _____

_____

_____

_____

11. Give the approximate (a) Hubble age of the observable universe _____
_____ ; (b) Hubble radius _____ .

Compare your answers to the questions on the self-test with the answers given below. If all of your answers are correct, you are ready to go on to the next chapter. If you missed any questions, review the sections indicated in parentheses following the answer. If you missed several questions, you should probably reread the entire chapter carefully.

1. Cosmology is the branch of science concerned with the origin, present structure, evolution, and final destiny of the universe.   (Section 7.1)

2. Cosmological models do not give a supernatural cause or meaning to physical events, but try to explain these events using only the laws of nature and mathematics.   (Section 7.1)

3. Light from distant galaxies is shifted in wavelength toward the red end of the spectrum, a phenomenon called the cosmological redshift. The farther away a galaxy is, the greater its redshift. The most distant galaxies are receding uniformly from us and from each other.   (Section 7.2)

4. The Hubble law says that a galaxy's velocity of recession (v) is directly proportional to its distance away from us (d). The Hubble law can be written algebraically as

   $v = H_0 d$ where H_0 is called the Hubble constant.   (Section 7.4)

5. The Hubble constant is very important because it gives the rate at which the galaxies are receding, or the rate at which the universe is expanding. It is a basis for estimating the size and age of the universe.   (Sections 7.4, 7.10, 7.15, 7.16)

6. The cosmological principle states that on a sufficiently large scale, the universe is homogeneous and isotropic at any given time.   (Section 7.5)

7. (a) 1, 2, and 3;   (b) 2 and 3;   (c) 1, 2, and 3;   (d) 1 and 3;   (e) 2.   (Sections 7.2, 7.6, 7.7, 7.10, 7.11, 7.14)

8. (1) Rate of change of the Hubble constant with time;   (2) density of matter and energy in the universe.   (Sections 7.10, 7.11)

9. The microwave radiation striking Earth from all directions in space provides strong evidence for the Big Bang model. It appears to be the redshifted remnant of the radiation created in the Big Bang.   (Section 7.12)

10. A brief phase of incredibly rapid expansion shortly after the Big Bang could explain how an explosive beginning could result in both the homogeneity of the cosmic microwave background and the large-scale structure of the observable universe.   (Section 7.14)

11. (a) 13.8 billion years; (b) about 14 billion light-years. (Sections 7.15, 7.16)

8

EXPLORING THE SOLAR SYSTEM

Finally we shall place the Sun himself at the center of the Universe.

Nicolaus Copernicus

De Revolutionibus Orbium Coelestium (1543)

Objectives

☆ List the members of the solar system.

☆ State the essential difference between a planet and a star.

☆ Describe evidence supporting the nebular theory of the formation of the solar system.

☆ Explain the phases of the Moon.

☆ Outline the development of our understanding of the solar system, including the contributions of Ptolemy, Copernicus, Galileo, Tycho Brahe, Kepler, and Newton.

☆ State and apply the laws governing the motions of bodies under gravity.

☆ Explain the apparent motions of the planets, including retrograde motion.

☆ Explain why the Moon's sidereal month and synodic month differ.

☆ Differentiate between the revolution and rotation of celestial bodies.

☆ Explain the motions of Earth-orbiting satellites and interplanetary spacecraft.

☆ Compare and contrast the general properties of the eight major planets and their moons.

☆ Describe the asteroids (minor planets) and the Kuiper Belt objects.

8.1 INVENTORY

Our **solar system** includes the Sun and all objects gravitationally bound to it. These objects are planets, dwarf planets, moons, and small solar system bodies—asteroids (also called minor planets), comets, Kuiper Belt objects, and interplanetary dust and gas. ▶photojournal.jpl.nasa.gov◀

A solar system **planet** is a body that orbits the Sun and has sufficient mass/gravity to be nearly round and to clear the region around its orbit. The planets range in mass and size from lightest, smallest Mercury to heaviest, largest Jupiter. All of the planets together are only 0.001 as massive and 0.3 as big as the Sun (Figure 8.1).

A **dwarf planet** orbits the Sun, has sufficient mass/gravity to be nearly round but not to clear its neighborhood, and is not a satellite.

Stars generate their own light, but planets cannot because they are much less massive and colder. They shine by reflecting starlight.

What is the essential difference between a planet and a star? _____

Answer: Mass. A planet is much less massive and colder than a star. It shines by reflecting light from a star. (A star generates its own light internally by nuclear fusion.)

Figure 8.1. The major planets of the solar system. Largest Jupiter is over 11 times the size of Earth.

8.2 ORIGIN OF THE SOLAR SYSTEM

The **solar nebular model** says that the solar system formed out of an eastward rotating interstellar cloud about 4.6 billion years ago (Section 4.3). The nebula contracted into the proto-Sun surrounded by a spinning disk where the planets formed as dust and gas accreted (Figure 4.4). The new Sun blew away most residual gas and dust.

The nebular theory is supported by the properties of the solar system observed today.

All of the planets **revolve**, or travel around, the Sun in the same direction as their originator clouds, or counterclockwise as seen from above. Most moons orbit the same way. This movement is called **direct motion** (Figure 8.2). The planets rotate as they revolve. Their rotation (except for Venus and Uranus) is also direct.

The mean plane of Earth's orbit around the Sun is called the **ecliptic**. The orbits of all of the planets are in nearly this same plane, like the lanes on a running track.

The planets whose orbits are closer to the Sun than Earth's are called **inferior**, while those whose orbits are outside Earth's are called **superior**.

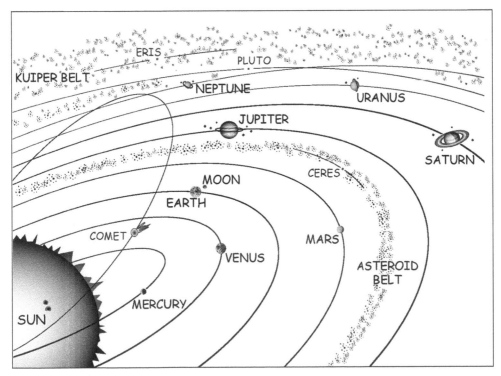

Figure 8.2. Our solar system (not to scale).

Refer to Figure 8.2. List the (a) inferior and (b) superior planets.

(a) _____

(b) _____

_____

Answer: (a) Mercury, Venus; (b) Mars, Jupiter, Saturn, Uranus, Neptune.

8.3 DAY NAMES

Five planets—Mercury, Venus, Mars, Jupiter, and Saturn—look like very bright stars in the sky. The Sun, Moon, and these five bright planets were known to the ancients. Each was thought to rule one day of the week, which was given its name (in Latin).

We take the names for our days of the week from the Anglo-Saxons, who substituted the names of equivalent gods and goddesses for the Roman ones. French and Spanish names are adapted directly from the Latin.

Refer to Table 8.1. (a) Which of our days of the week is closest to the original Latin god's name? _____ (b) Which days of the week carry the names of the Anglo-Saxon equivalent? _____

_____

Answer: (a) Saturday; (b) Monday, Tuesday, Wednesday, Thursday, Friday.

TABLE 8.1 Days of the Week

Day	Ruling Planet	Anglo-Saxon Equivalent	Latin	French	Spanish
Sunday	Sun	—	*Dies Solis*	*Dimanche*	*Domingo*
Monday	Moon	Mona	*Dies Lunae*	*Lundi*	*Lunes*
Tuesday	Mars	Tiw	*Dies Martis*	*Mardi*	*Martes*
Wednesday	Mercury	Woden	*Dies Mercurii*	*Mercredi*	*Miercoles*
Thursday	Jupiter (Jove)	Thor	*Dies Jovis*	*Jeudi*	*Jueves*
Friday	Venus	Frigg	*Dies Veneris*	*Vendredi*	*Viernes*
Saturday	Saturn	Seterne	*Dies Saturni*	*Samedi*	*Sabado*

8.4 MOON PHASES

The Moon is Earth's only natural satellite, orbiting our planet as we travel around the Sun. It shines in the sky by reflecting sunlight.

The Moon's appearance changes regularly every month. Half of the Moon is always lighted by the Sun, but the bright shape we see from Earth, called its **phase**, changes as the Moon travels around our planet. The recurring cycle of apparent shapes is called the **phases of the Moon**.

Refer to Figure 8.3. The **new Moon** is dark. It is not seen in the sky because the Moon's dark side is facing Earth. The **waxing** (growing bigger) **crescent** Moon follows a few days later. You often see its disk faintly lighted by sunlight reflected from Earth, called **earthshine**.

About 7 days after the new Moon, when the Moon has traveled ¼ of its way around Earth, it rises around noon, and we observe **first quarter** shine. The **waxing gibbous** Moon follows, with more than half of the Moon's bright disk shining toward Earth.

When the Moon is about two weeks into its cycle, the **full Moon** lights the sky all night with its whole, bright disk. Full Moon occurs 12.37 times a year. So

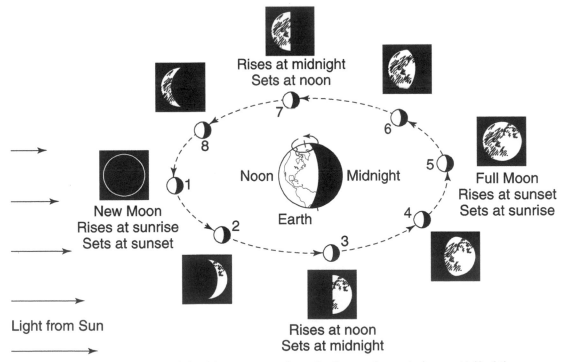

Figure 8.3. The phases of the Moon, as seen from Earth, are shown in boxes. Half of the Moon is always lighted by the Sun, as shown in the inner circle. The Moon's elongation is counted eastward around the sky. Elongations of 0°, 90°, 180°, 270° correspond to new, first quarter, full, and last quarter Moon, respectively.

months having two full Moons occur every 2.72 years on the average. The second full Moon within a particular month is called **blue Moon**. ("Once in a blue moon" means "very seldom.") Approximately once every 19 years, a year will have two months with two full Moons because February will not have a full Moon.

The visible part of the Moon's bright disk **wanes**, or decreases in size, as the Moon completes its trip around Earth during the last two weeks of its cycle.

The mean time required for the Moon's phases to repeat, called a **synodic month** or **lunation**, is 29.5 days. In this context, the age of the Moon is the time since new Moon. ▶http://apod.nasa.gov/apod/ap120205.html◀

Identify the phase of the Moon corresponding to the indicated position in its orbit for each of the following phases, as labeled in Figure 8.3: (a) waxing crescent _____ ; (b) first quarter _____ ; (c) waxing gibbous _____ ; (d) waning gibbous _____ ; (e) third quarter _____ ; (f) waning crescent _____ .

Answer: (a) 2; (b) 3; (c) 4; (d) 6; (e) 7; (f) 8.

Observe the Moon daily for a month, if possible. Keep a record of its bright shape, its position relative to the Sun, and its time of rising and setting. Use the information in this chapter to explain the changes you observe.

8.5 PLANET WATCHING

Planet positions are not marked on star maps. Stars keep their same relative positions in the sky for decades, but planets do not. The word "planet" comes from the Greek for "wanderer." Bright planets move across the celestial sphere near the ecliptic.

Venus and Mercury appear to move forward and backward on either side of the Sun in Earth's sky. Maximum **elongation**, or distance east or west of the Sun, is 48° for Venus and 28° for Mercury.

Mars, Jupiter, and Saturn wander generally eastward through the constellations of the zodiac. At times planets seem to reverse and move westward, called **retrograde motion**, before resuming direct motion. The apparent backward swing plus resumed forward motion is called a **retrograde loop** (Figure 8.4).

You can look up the exact locations of the planets on any given night in astronomical publications, websites, software, and apps (Useful Resources).

Figure 8.4. Successive observed positions of Mars among the zodiac constellations during a year.

Suggest a way of finding a particular planet such as Jupiter in the sky tonight.

_____

_____

_____

_____

_____

_____

Answer: Find out which zodiac constellation Jupiter is in today by using an astronomical publication, website, software, or an app. Locate that constellation on your star maps. For example, suppose you find Jupiter is in Taurus. If Taurus is in the sky tonight, you can easily spot Jupiter. It will be the brilliant "star" that does not belong to the constellation (Figure 8.5).

Figure 8.5. Jupiter in the constellation Taurus.

★ Keep a record of the position of Venus and Mars for several months. Observe them in the sky, if possible. Use the information in this chapter to explain the motions you observe. Tips for amateurs can be found at Association of Lunar and Planetary Observers. ▶http://alpo-astronomy.org◀

8.6 BRIEF HISTORY

The search for a simple explanation of the planets' observed motions in the sky changed humanity's view of the world.

In *Almagest*, written about A.D. 150, the Alexandrian astronomer Ptolemy (Claudius Ptolemaeus) described the ancient **geocentric**, or Earth-centered, **model** of the universe. Circles were considered "perfect" shapes. The Sun, Moon, and planets were supposed to move on small circles called **epicycles**, whose centers moved around Earth on larger circles called **deferents**.

For more than 14 centuries, this **Ptolemaic system** was accepted as the basis for astronomical work. It described with considerable accuracy the observed positions and motions of the heavenly bodies known at that time. It also expressed the view of the world that people had from observing the sky. With minor modifications, this Earth-centered theory became part of the dogma of the Roman Catholic Church of the Middle Ages.

Polish astronomer Nicolaus Copernicus (1473–1543) published his radical **heliocentric**, or Sun-centered, **model** the year he died. In the **Copernican system**, the planets, including Earth, circle around a stationary central Sun. According to the Copernican model, the apparent wandering motions of the planets result from a combination of the real orbital motions of both Earth and observed planets.

The apparent motion of Mars illustrated in Figure 8.6 is explained as follows: Mars never really moves backward in its orbit. Planets always go forward. The retrograde loop in the sky is caused by the relative motion of Earth and Mars. Our faster-moving Earth catches up to Mars and moves past it. So Mars, the outer planet, looks to us as if it is moving backward. It is like watching a slower car while you are overtaking it in a faster car.

What change in the philosophical view of Earth was required for people to accept the Copernican theory instead of the Ptolemaic? _____

_____

Answer: Earth could no longer be considered the center of the entire universe and supremely important or unique.

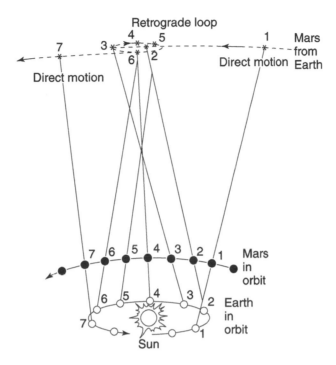

Figure 8.6. The apparent motion of Mars explained by the heliocentric model. Numbers mark positions at one-month intervals. (1–3): Mars appears to slow its direct motion as Earth overtakes it. (4): Mars appears to move backward as Earth passes it. (5–7): Mars resumes direct motion as Earth moves ahead of it.

8.7 FIRST TELESCOPIC DATA

Italian scientist Galileo Galilei (1564–1642) was the first person to use a telescope for scientific sky observations, in 1609. His observations provided important data for the Copernican theory.

The sight of mountains, craters, and extensive dark areas on the Moon, as well as sunspots and their movements, convinced Galileo that the heavens were not perfect and unchanging. His discovery of four large moons orbiting Jupiter confirmed that Earth was not the center of all heavenly motions.

Galileo observed that bright Venus appears to change its shape and size regularly. The Ptolemaic system could not account for the phases of Venus. But the Copernican system had a simple explanation.

Refer to Figure 8.7. Venus and Mercury, the two inferior planets, show phases as they reflect sunlight to Earth from different places in their orbits around the Sun. An inferior planet looks most fully lit in its gibbous phase, near **superior conjunction**, the point on the far side of the Sun from Earth.

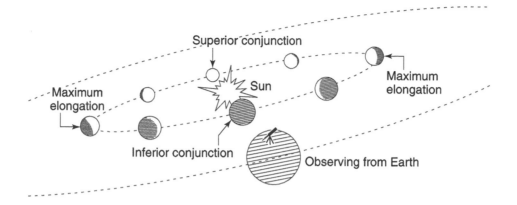

Figure 8.7. The phases of Venus as viewed through a telescope from Earth.

It appears as a crescent and biggest near **inferior conjunction**, the point between Earth and Sun.

The Roman Catholic Church in 1616 banned books that supported Copernicanism. Galileo was allowed to continue research, provided he did not hold, teach, or defend heretical doctrines. But his *Dialogue Concerning the Two Chief World Systems*, published in 1632, supported the Copernican system.

The next year, Galileo, almost 70 years old and after a brilliant career, was forced by the Inquisition to recant his astronomical findings and condemned to house arrest. Debates about the compatibility of religion and science still rage. The Church slowly acted to rectify the infamous verdict and finally vindicated Galileo in 1992. Enjoy a virtual visit to the Museo Galileo in Florence, Italy, and see Galileo artifacts. ▶http://www.museogalileo.it/en◀

8.8 LAWS OF PLANETARY MOTION

German astronomer Johannes Kepler (1571–1630) deduced a simple, precise description of planetary motion. Kepler worked from records inherited from Tycho Brahe (1546–1601), a Danish astronomer who had recorded the positions of stars and planets with unprecedented accuracy for almost 20 years. The remains of Tycho's observatory and artifacts are on display on the now-Swedish island of Ven.

Kepler's laws of planetary motion greatly improved the accuracy of predictions of planet positions. The three laws state:

1. Each planet moves around the Sun in an orbit that is an ellipse with the Sun at one focus.

2. Each planet moves so that an imaginary line joining the Sun and the

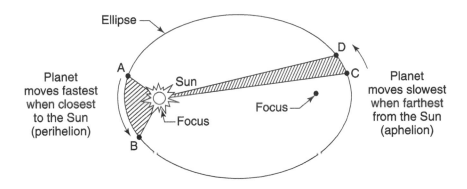

Figure 8.8. Planetary motion.

planet sweeps out equal areas in equal times. As illustrated in Figure 8.8, the planet goes from A to B and C to D in the same amount of time. In other words, planets move fastest when they are closest to the Sun (**perihelion**) and slowest when they are farthest away (**aphelion**).

3. The squares of the periods of time required for any two planets to complete a trip around the Sun have the same ratio as the cubes of their average distances from the Sun.

Kepler's third law can be used to find a planet's average distance, d, from the Sun compared to Earth's average distance of 1 AU (Section 4.2). The planet's orbital period, p, in years is found from observations. Then Kepler's third law is written $d^3 = p^2$.

For example, Jupiter's orbital period is 11.86 years. So Jupiter's average distance, d, from the Sun is found from $d^3 = (11.86)^2 \cong 141$. Solving for d = $\sqrt[3]{141}$, we find d = 5.2 AU.

How far would a planet be from the Sun if its orbital period were observed to be 8 years? Explain. _____

_____

Answer: 4 AU, or 4 times Earth's average distance. According to Kepler's third law, $d^3 = p^2$. So $d^3 = (8)^2 = 64$. And d = $\sqrt[3]{64} = 4$.

Note: An **ellipse** is a closed curve for which the sum of the distances from any point on the curve to two fixed points, or **foci**, inside is a constant. The **semi-major axis**, half the maximum dimension (major axis), defines size. **Eccentricity**, or distance between foci divided by major axis, measures how much the ellipse deviates from being a circle.

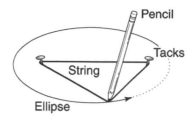

Figure 8.9. Drawing an ellipse.

To draw an ellipse, put two tacks for foci in a board. Tie a string around them. Trace the ellipse by keeping the string taut with your pencil point (Figure 8.9).

8.9 MOTION AND GRAVITY

Kepler's laws explain how the planets are observed to move. English physicist and mathematician Sir Isaac Newton (1642–1727) formulated laws that explain why the planets move as they do. His book *The Mathematical Principles of Natural Philosophy* was published in 1687.

Newton's laws of motion state:

1. A body stays in a state of rest or uniform motion unless an outside force acts on it.

2. The net force, F, on a body is equal to its mass, m, multiplied by its acceleration, a. The formula is: F = ma.

3. Whenever one body exerts a force on a second body, the second body exerts an equal and opposite force on the first body.

Newton's **law of gravity** states: Any two objects of masses m_1 and m_2 separated from each other by a distance, d, attract each other with a force, F, called gravity, that is directly proportional to the product of their masses and inversely proportional to the square of their distance away from each other (Figure 8.10). The formula is:

$$F = \frac{G\, m_1 m_2}{d^2}$$ where G = gravitational constant (Appendix 2).

Figure 8.10. Newton's law of gravity.

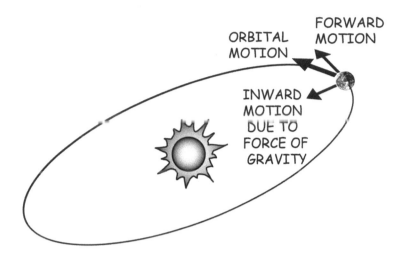

Figure 8.11. The combination of two motions that keep a planet traveling in orbit.

A force of attraction is necessary to keep the planets moving in their curved paths around the Sun. Without this force they would move straight away into space. The required force is provided by the Sun's gravity, which continuously pulls the planets in toward the Sun.

The combination of their forward motion and their motion in toward the Sun under gravity keeps the planet traveling in its orbit around the Sun (Figure 8.11).

Newton's genius was to realize that his law of gravity applies to falling objects on Earth, to the motion of the Moon and planets, and to all material bodies. He proposed that his law of gravitation and his three laws of motion were basic laws of physics. His laws are **universal**, true for all objects everywhere in the universe.

Newton generalized and mathematically derived Kepler's laws of planetary motion from basic principles. He invented and used in his work the branch of mathematics we call calculus.

Apply Newton's laws to explain why moons stay in orbit around their parent planets._____

_____

_____

Answer: A combination of two motions keeps a moon in orbit around its parent planet—its forward motion and its inward motion caused by the pull of the planet's gravity.

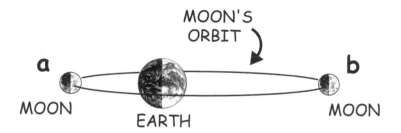

Figure 8.12. The Moon's orbit around Earth.

8.10 MOON'S ORBITAL MOTION

The Moon's orbit is an ellipse with Earth at one focus. Its travels at an average speed of 1.02 km/sec (2295 miles per hour).

The time for the Moon to complete one trip around Earth with respect to the stars, about 27.3 days, is called a **sidereal month**.

The Moon's average angular diameter in the sky is about ½° (31'5" of arc). The Moon looks larger at **perigee**, the point in its orbit closest to Earth, and smaller at **apogee**, the point in its orbit farthest from Earth.

Identify the apogee and perigee in Figure 8.12, and indicate where the Moon looks larger than average and where it looks smaller than average. (a) _____ _____ ; (b) _____

Answer: (a) Perigee—the Moon looks larger; (b) apogee—the Moon looks smaller.

8.11 SYZYGY

If you enjoy word games or crossword puzzles, you'll find **syzygy** a good word to know. It means three celestial bodies in a line, such as Sun-Moon-Earth.

Refer to Figure 8.13. Explain why a synodic month, or the month of the Moon's phases, is 2 days longer than a sidereal month. _____

_____

_____

_____

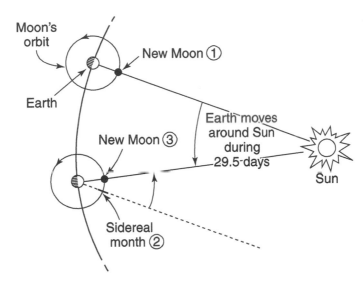

Figure 8.13. The interval from one new Moon to the next is 2 days longer than a sidereal month.

Answer: Start with new Moon (1). After 27.3 days, the Moon has traveled completely around Earth (2). But Earth and Moon have also moved together around the Sun during that time. Two more days must pass before the Moon, Earth, and Sun are lined up so that the Moon is new again (3).

8.12 SPACEFLIGHT

Spacecraft obey the same basic laws of physics that natural astronomical bodies do.

Any body in orbit around a larger parent body is called a **satellite**. Rockets launch artificial satellites into Earth-orbit with a forward velocity of at least 8 km/sec (17,300 miles per hour). The combination of their forward motion and their motion in toward Earth under Earth's gravity keeps the satellites in their orbits. Most are designed to burn up due to friction if they plunge back into the atmosphere. Piloted craft and selected payloads are made to survive re-entry and land safely.

Russian Sputnik 1, an 82-kg (180-pound) metal ball with a transmitter and batteries was the first artificial satellite. Rocketed to space on October 4, 1957, Sputnik 1 signaled the opening of the space age. Today hundreds of robot communications, weather, research, navigation, and military satellites of many nations are operating in Earth orbits.

Robot spacecraft are sent to explore the planets. These spacecraft are launched with a forward velocity into orbit around the Sun. Their motions are calculated using Newton's laws, just as planetary motions are.

All of the major planets have been observed close up by robot spacecraft. These planet probes are packed with cameras, data sensors, and computers programmed to operate automatically far from direct human interaction. None has returned so far. They radio images and data back to Earth for analysis. ▶http://solarsystem.nasa.gov/missions◀

The most ambitious multi-planet space flight yet was the U.S. Project Voyager. Twins Voyager 1 and Voyager 2 (Figure 8.14), were launched in 1977 to take advantage of a four-planet lineup that occurs once in 176 years.

Down-to-the-minute trajectory planning was done. **Gravity assist**, a technique of using a planet's gravitational field to change a spacecraft's velocity without consuming fuel, was used at each encounter to increase Voyager's speed and bend its flight path enough to convey it to the next destination.

Voyagers 1 and 2 returned a total of 118,000 images and data that revolutionized the science of planetary astronomy. Now, in an extended mission,

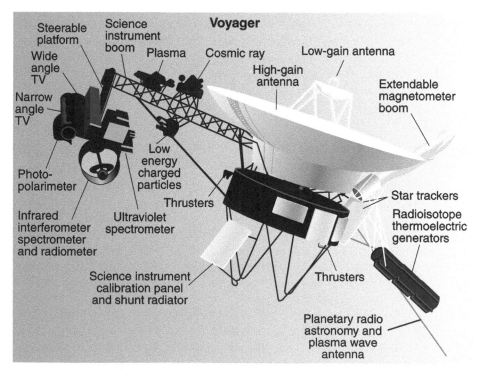

Figure 8.14. The 800-kg (nearly 1 ton) Voyager spacecraft. It has 11 sets of target body (point at object) and particles, fields, and waves sensors. Nuclear-electric generators power the spacecraft instruments, radio, and computers.

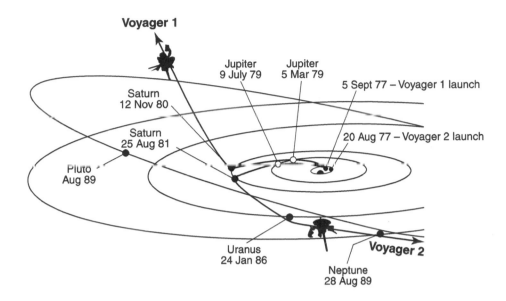

Figure 8.15. Project Voyager timetable. The 12-year primary mission included encounters with Jupiter, Saturn, Uranus, and Neptune and their ring systems and moons. Voyager 2 was launched first. Voyager 1 was launched 16 days later on a faster, shorter trajectory. Encounter dates are marked where spacecraft and planet paths intersected.

Voyager Interstellar Mission (VIM), Voyager 1 is escaping the solar system 35° above the ecliptic plane at a rate of about 3.6 AU per year. Voyager 2 is speeding out 48° below the ecliptic plane at some 3.3 AU per year.
►http://voyager.jpl.nasa.gov◄

Refer to Figure 8.15. About how many years after launch did Voyager 2 reach
(a) Jupiter? _____ (b) Saturn? _____ (c) Uranus? _____
(d) Neptune? _____

Answer: (a) 2 years; (b) 4 years; (c) 8.5 years; (d) 12 years.

8.13 PLANET SURVEY

Comparative planetology, the study of one planet compared with others, helps us to better understand our own world as well as the rest of our solar system. The general properties of the eight planets are listed in Table 8.2.

TABLE 8.2 Properties of the Planets

	Mercury	Venus	Earth
Mean distance from Sun			
millions km	57.9	108.2	149.6
(millions of miles)	(36)	(68)	(93)
Astronomical Units, AU	0.39	0.72	1.00
Mean orbital velocity, km/sec	47.87	35.02	29.79
Period of revolution, sidereal	87.97 days	224.70 days	365.26 days
Synodic (days)	116	584	—
Rotation period, sidereal (*days*, *hours*, *minutes*, *seconds*)	58.646d	243.019d	23h 56m 4s
Inclination of orbit to ecliptic	7°00′	3°23′	0°00′
Eccentricity of orbit	0.206	0.007	0.017
Oblateness	0	0	0.0034
Equatorial diameter, km	4,879	12,104	12,756
(miles)	(3,030)	(7,520)	(7,930)
Mass (Earth = 1)	0.06	0.82	1.00
Density, t/m^3	5.43	5.24	5.52
Surface gravity (Earth = 1)	0.38	0.90	1.00
Confirmed satellites	0	0	1 moon

Mercury, Venus, Earth, and Mars have similar physical and orbital characteristics. They are called **terrestrial**, or earthlike, **planets**. Jupiter, Saturn, Uranus, and Neptune are also similar to one another and are called **giant**, or **Jovian** (meaning Jupiter-like), planets.

Examine Table 8.2. How do the terrestrial planets differ from the giant planets in (a) distance from the Sun? (b) size? (c) mass? (d) density?

Terrestrial Planets

(a) _____

(b) _____

(c) _____

(d) _____

Giant Planets

(a) _____

(b) _____

(c) _____

(d) _____

TABLE 8.2 Properties of the Planets (*Continued*)

Mars	Jupiter	Saturn	Uranus	Neptune
227.9	778.3	1429.4	2875.0	4504.4
(142)	(486)	(888)	(1787)	(2799)
1.52	5.20	9.56	19.22	30.11
24.13	13.07	9.69	6.80	5.43
686.98 days	11.86 years	29.46 years	84.01 years	164.79 years
780	399	378	370	367
24h 37m 26s	9.842h	10.656h	17.239h	16.109h
1°51′	1°18′	2°29′	0°46′	1°46′
0.093	0.048	0.056	0.046	0.009
0.0068	0.065	0.098	0.023	0.017
6,792	142,980	120,536	51,120	49,530
(4,220)	(88,850)	(74,900)	(31,770)	(30,780)
0.11	317.83	95.16	14.54	17.15
3.94	1.33	0.69	1.27	1.64
0.38	2.53	1.06	0.90	1.14
2 moons	63 moons	60 moons	27 moons	13 moons
	Rings	Rings	Rings	Rings

Answer:

	Terrestrial Planets	**Giant Planets**
(a)	Near the Sun.	Far from the Sun.
(b)	Small diameter.	Large diameter.
(c)	Small mass.	Large mass.
(d)	High density.	Low density.

8.14 DAYS AND YEARS

The **period of revolution** is the length of time for a celestial body to go around its orbit once.

A planet's **sidereal revolution period** is measured relative to the stars. It is the length of the planet's year in terms of Earth time. A planet's **synodic revolution period** is the planet's orbital period as seen from Earth. It is equal to the time for a planet to return to a specific **aspect**, or certain position relative to the Sun as seen from Earth, such as conjunction.

A planet's synodic period differs from its sidereal period because Earth is itself moving in its orbit around the Sun.

The **period of rotation** is the length of time required for a celestial body to turn around on its axis once. A planet's **sidereal rotation period** is the length of one sidereal day on the planet (Section 1.23). A planet's **synodic rotation period** is the length of one solar day on the planet. It is the time interval between two successive meridian transits of the Sun as would be seen by an observer on that planet.

Earth's rotation was long used for timekeeping. But Earth's rotation is not exactly uniform. Its accuracy is good to about 0.001 second a day. Atomic clocks, which operate by measuring the resonant frequency of a given atom—cesium, hydrogen, or mercury—are accurate to a billionth of a second a day. The International Earth Rotation Service monitors the difference in the two time scales and adds one-second steps, called **leap seconds**, to the world's clocks as necessary.

Examine Table 8.2. (a) Which giant planet has the longest year, and how many Earth-years is it equal to? (b) Which planet has the longest sidereal day, and how many Earth-days is it equal to? _____

_____

_____

Answer: (a) Neptune. It is equal to 164.8 Earth-years. (b) Venus. It is equal to 243 Earth-days.

8.15 ASTEROIDS

Asteroids (from the Greek for "starlike"), or minor planets, look like stars in a telescope. Most are irregularly shaped, cratered, rotating chunks of rock and/or metal covered by rubble and soil.

Millions orbit the Sun in the **asteroid belt**, a region from 2.0 to 3.3 AU away between the orbits of Mars and Jupiter. Some have moons. Their total mass is less than our Moon's.

The largest object was discovered first, by Sicilian astronomer Giuseppi Piazzi (1746–1826) in 1801. Numbered 1 and named Ceres, it is a dwarf planet 950 km (590 miles) across. Some 600,000 asteroids are numbered and 16,000 are named also. More, down to about 10m, are added each year.

Robot spacecraft explore asteroids. Record firsts include U.S. Galileo / close-ups of 951 Gaspra, in 1991, (Figure 8.16) and small moon Dactyl, orbiting 243 Ida, in 1993; U.S. Near Shoemaker / data from a landing on 433 Erosin 2001; JAXA Hayabusa/return of samples of surface material for analysis from 25,143 Itokawa, in 2010; U.S. Dawn / mineral composition, topo-

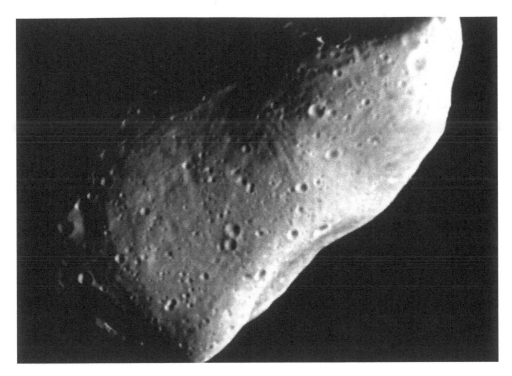

Figure 8.16. First closeup of asteroid 951 Gaspra, from U.S. robot Galileo when they were only 16,000 km apart. Gaspra measures about 20 by 12 by 11 km. The smallest craters seen are some 300 m across.

graphical, and geological data from year-long orbit of 4 Vesta, in 2011–2012, plus close look at Ceres, in 2015. ▶http://solarsystem.nasa.gov/planets/profile.cfm?Object=Asteroids◀

Asteroids are classified by how they reflect sunlight using **sp to -tometry**, the accurate determination of magnitudes within specified wavelength regions. Very dark **C-type asteroids**, so called because they are carbonaceous, are very common in the outer asteroid belt. Moderately bright **S-type asteroids** contain silicates mixed with metals. They are common in the inner belt. Very bright **M-type asteroids** are metallic.

The bright asteroids are probably clumps of mass that condensed from the original solar nebula but never got big enough to form a large planet. The brightest, 4 Vesta, is 510 km (320 miles) in diameter. The fainter ones are probably fragments resulting from numerous collisions.

Near-Earth objects (NEOs) are asteroids and comets that regularly come nearby. **Aten asteroids** have orbits inside Earth's. **Apollo asteroids** cross and pass inside Earth's orbit to their perihelion. Some have come within a million kilometers of us. **Amor asteroids**, with orbits between 1 and 1.3 AU, stay beyond Earth's orbit.

TABLE 8.3 Selected Moons of the Planets

All but three of the known moons orbit giant planets at distances that suggest their origins. Ring moons, up to 200 km in diameter, are closest. Regular moons circle in or near the planets' equatorial planes beyond three planetary radii. Irregular bodies up to 200 km in diameter mostly go in eccentric orbits at large inclinations to the planets' orbital planes. ▶http://solarsystem.nasa.gov/planets/index.cfm◀

Planet	Satellite	Diameter (km)	Mean Distance from Planet[a] (10^3 km)	Revolution Period[b] (days)	Discovery
Earth	Moon	3,475	384.4	27.322	
Mars	Phobos	22	9.4	0.319	Hall, 1877
	Deimos	12	23.5	1.262	Hall, 1877
Jupiter	Metis	43	128.0	0.3	Voyager 2, 1979
	Adrastea	16	129.0	0.3	Jewett, Danielson,1979
	Amalthea	167	181.4	0.5	Barnard, 1892
	Thebe	99	221.9	0.7	Voyager 1, 1979
	Io	3,643	421.8	1.8	Galileo, 1610
	Europa	3,122	671.1	3.6	Galileo, 1610
	Ganymede	5,262	1,070.	7.2	Galileo, 1610
	Callisto	4,820	1,883.	16.7	Galileo, 1610
	Themisto	4	7,284.	130.0	Kowal, 1975
	Leda	20	11,165.	241.0	Kowal, 1974
	Himalia	185	11,461.	251.0	Perrine, 1904
	Lysithea	35	11,717.	259.0	Nicholson, 1938
	Elara	85	11,741.	260.0	Perrine, 1905
	Carpo	3	16,989.	456.0	Sheppard et al., 2003
	Orthosie	5	20,720.	632 R	Sheppard et al., 2000
	Euanthe	3	20,797.	620 R	Sheppard et al., 2001
	Praxidike	5	20,907.	625 R	Sheppard et al., 2000
	Iocaste	4	20,061.	623 R	Sheppard et al., 2001
	Mneme	2	21,069	620 R	Gladman et al., 2003
	Hermippe	4	21,131.	634 R	Sheppard et al., 2001
	Thelxinoe	2	21,162	628 R	Sheppard et al., 2004
	Ananke	30	21,276.	630 R	Nicholson, 1951
	Arche	3	22,931	724 R	Sheppard et al., 2002
	Pasithee	2	23,004	719 R	Sheppard et al., 2001
	Isonoe	3	23,155.	726 R	Sheppard et al., 2000
	Erinome	3	23,196.	729 R	Sheppard et al., 2000
	Kale	2	23,217	729 R	Sheppard et al., 2001
	Aitne	3	23,229.	730 R	Sheppard et al., 2001
	Taygete	4	23,280	732 R	Sheppard et al., 2000
	Carme	45	23,404.	734 R	Nicholson, 1938
	Pasiphae	60	23,624.	744 R	Melotte, 1908
	Eukelade	4	23,661.	746 R	Sheppard et al., 2003
	Hegemone	3	23,947.	740 R	Sheppard et al., 2003
	Sinope	38	23,939.	759 R	Nicholson, 1914
	Callirrhoe	7	24,103	759 R	Scotti, Spahr 1999
Saturn	Pan	26	133.6	0.6	Showalter, 1990
	Daphnis	7	136.5	0.6	Cassini ISS, 2005
	Prometheus	94	139.4	0.6	Voyager 1, 1980
	Pandora	81	141.7	0.6	Voyager 1, 1980
	Janus*	180	151.5	0.7	Dollfus, 1966
	Epimetheus*	120	151.4	0.7	Voyager 1,1980
	Mimas	397	185.6	0.9	Herschel, 1789
	Enceladus	499	238.1	1.4	Herschel, 1789

Planet	Satellite	Diameter (km)	Mean Distance from Planet[a] (10³ km)	Revolution Period[b] (days)	Discovery
	Tethys	1,060	294.7	1.9	Cassini, 1684
	Telesto	24	294.7	1.9	Voyager 1, 1980
	Calypso	19	294.7	1.9	Voyager 1, 1980
	Dione	1,118	377.4	2.7	Cassini, 1684
	Helene	32	377.4	2.7	Laques, Lecacheux, 1980
	Rhea	1,528	527.1	4.5	Cassini, 1672
	Titan	5,150	1,221.9	16.0	Huygens, 1655
	Hyperion	283	1,464.1	21.3	Bond, Lassell, 1848
	Iapetus	1,436	3,560.8	79.3	Cassini, 1671
	Kiviuq	14	11,365.	449	Gladman, 2000
	Ijiraq	10	11,440.	451	Kavelaars et al., 2000
	Phoebe	220	12,944.3	550.4 R	Pickering, 1898
	Paaliaq	19	15,199.	687	Gladman, 2000
	Skadi	6	15,647.	729 R	Kavelaars et al., 2000
	Albiorix	26	16,404.	783	Holman, Spahr, 2000
	Erriapus	9	17,616.	872	Kavelaars et al., 2000
	Siarnaq	32	18,160.	893	Gladman et al., 2000
	Tarvos	13	18,247.	926	Kavelaars et al., 2000
	Mundilfari	6	18,709.	951 R	Gladman et al., 2000
	Suttung	6	19,463.	1010 R	Gladman et al., 2000
	Thrym	6	20,382.	1087 R	Gladman et al., 2000
	Ymir	16	23,096.	1312 R	Gladman, 2000
Uranus	Cordelia	40	49.8	0.4	Voyager 2, 1986
	Ophelia	23	53.8	0.4	Voyager 2, 1986
	Bianca	51	59.2	0.4	Voyager 2, 1986
	Cressida	80	61.8	0.5	Voyager 2, 1986
	Desdemona	64	62.7	0.5	Voyager 2, 1986
	Juliet	94	64.4	0.5	Voyager 2, 1986
	Portia	135	66.1	0.5	Voyager 2, 1986
	Rosalind	72	69.9	0.6	Voyager 2, 1986
	Belinda	81	75.3	0.6	Voyager 2, 1986
	Puck	162	86.	0.8	Voyager 2, 1985
	Miranda	472	129.9	1.4	Kuiper, 1948
	Ariel	1,158	190.9	2.5	Lassell, 1851
	Umbriel	1,169	266.	4.1	Lassell, 1851
	Titania	1,578	436.3	8.706	Herschel, 1787
	Oberon	1,523	583.5	13.46	Herschel, 1787
	Caliban	98	7,231.	579.7 R	Gladman et al., 1997
	Stephano	20	8,004.	677 R	Gladman et al., 1999
	Trinculo	10	8,578.	759 R	Holman et al., 2001
	Sycorax	190	12,179.	1288. R	Nicholson et al., 1997
	Prospero	30	16,243.	1977 R	Holman et al., 1999
	Setebos	30	17,501.	2235 R	Kavelaars et al., 1999
Neptune	Naiad	58	48.2	.3	Voyager 2, 1989
	Thalassa	80	50.1	.3	Voyager 2, 1989
	Despina	148	52.5	.3	Voyager 2, 1989
	Larissa	192	73.5	.6	Voyager 2, 1989
	Proteus	416	117.6	1.1	Voyager 2, 1989
	Triton	2,707	354.8	5.877 R	Lassell, 1846
	Nereid	340	5 513.4	360.1	Kuiper, 1949
	Halimede	50	15,686	1875	Holman et al., 2002
	Psamathe	36	46,738	9136	Jewitt et al., 2003

[a]At mean opposition distance. [b]Sidereal period. R = Retrograde orbit. *Coorbital satellites.

When a nearby asteroid is first sighted, many people fear a disastrous collision. Astronomers judge asteroids that are larger than a kilometer in diameter as the greatest threat. Probably over a million exist. Modern telescopes could probably spot such a large asteroid decades before it reached Earth so a catastrophe could be prevented.

U.S. robot Dawn (2007–) is about to orbit and explore Vesta in 2011 and then Ceres in 2012. ▶http://dawn.jpl.nasa.gov◀ Water, in the form of water of hydration, was first detected on 1 Ceres. Metallic asteroids near Earth could be mined in the twenty-first century to provide raw materials for space colonists and interplanetary expeditions.

Kuiper Belt objects (KBOs) are small icy objects beyond Neptune. Some have moons. They orbit the Sun in the **Kuiper Belt,** a huge ring-shaped region from 30 AU to perhaps 150 AU near the ecliptic plane. It was named for Dutch-American Astronomer Gerard Kuiper (1905–1973), who predicted that KBOs exist from the theory of planet formation in 1951. KBOs likely are primordial leftovers that never coalesced into planets. Since the first KBO was sighted in 1992, hundreds more have been discovered.

What are asteroids? _____

_____

Answer: Swarms of irregular, rocky bodies that orbit the Sun, mostly between the orbits of Mars and Jupiter.

8.16 MOON COMPARISONS

Giant planets are much more massive with stronger gravity than terrestrial planets. Hence they can more readily hold moons that formed or passed nearby.

Refer to Tables 8.2 and 8.3. (a) How many satellites do terrestrial planets have altogether? (b) Why do the giant planets have many more satellites than do terrestrials? (c) List the moons of other planets that are larger than our Moon. (d) Which is the largest known moon in the solar system?

(a) _____

(b) _____

(c) _____

(d) _____

Answer: (a) Terrestrial planets have only three moons—Earth has one; Mars has two. (b) Giant planets have greater mass and so exert a stronger gravitational force. (c) Jupiter: Ganymede, Callisto, Io; Saturn: Titan. (d) Ganymede.

This self-test is designed to show you whether or not you have mastered the material in Chapter 8. Answer each question to the best of your ability. Correct answers and review instructions are given at the end of the test.

1. List the members of the solar system._____

_____

2. What is the essential difference between a planet and a dwarf planet? _____

_____

_____

3. Give two facts that support the nebular theory of the formation of the solar system._____

_____

4. In midnorthern latitudes, which phase of the Moon would you see if the Moon were rising in the sky about (a) 6:00 P.M.? _____

(b) noon? _____

5. Match each person to a contribution to the development of our understanding of the solar system.

____ (a) Described a geocentric view of the universe in the *Almagest* about A.D. 150.

____ (b) Determined his three laws of planetary motion empirically from observational data.

____ (c) First used a telescope for astronomical work and discovered the phases of Venus.

____ (d) Wrote a book describing a heliocentric model of planetary motions, which was published in 1543, the year he died.

____ (e) Formulated the three fundamental laws of motion and the universal law of gravitation.

____ (f) Observed and recorded planetary motions for almost 20 years.

(1) Copernicus.
(2) Galileo.
(3) Kepler.
(4) Newton.
(5) Ptolemy.
(6) Tycho Brahe.

6. What keeps planets in their orbits around the Sun? _____

_____

_____

7. Refer to Figure 8.17. Identify the following points: (a) Sun _____ ;
(b) ellipse _____ ; (c) aphelion _____ ; (d) perihelion _____
; (e) force of gravity is greatest _____ ; (f) planet that moves the
slowest _____ .

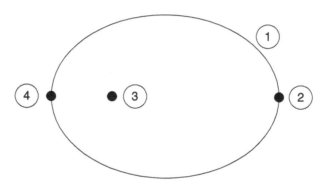

Figure 8.17. Planetary motion.

8. By how much do the Moon's sidereal month and synodic month differ?
Explain. _____

_____

_____

9. What force keeps spacecraft in their trajectories as they travel through the
solar system? _____

10. Classify each of the following as a property of (1) terrestrial planets or (2)
giant planets.

(a) Far from the Sun. _____

(b) Small diameter. _____

(c) Large mass. _____

(d) Low density. _____

(e) Short period of revolution. _____

(f) Short period of rotation. _____

(g) Many moons. _____

11. Match a planet to the appropriate description. *Hint*: Refer to Table 8.2.

____ (a)	Closest to Sun.	(1) Mercury.
____ (b)	Most dazzling rings.	(2) Venus.
____ (c)	Has the longest sidereal day.	(3) Earth.
____ (d)	Has a year approximately equal to 2 Earth-years.	(4) Mars.
		(5) Jupiter.
		(6) Saturn.
____ (e)	Most massive.	(7) Uranus.
____ (f)	Most dense.	(8) Neptune.

12. What are (1) asteroids and (2) Kuiper Belt Objects? _____

_____

_____

Help scientists learn more about our solar system. ▶http://cosmoquest.org/projects/vesta_mappers◀

ANSWERS

Compare your answers to the questions on the self-test with the answers given below. If all of your answers are correct, you are ready to go on to the next chapter. If you missed any questions, review the sections indicated in parentheses following the answer. If you missed several questions, you should probably reread the entire chapter carefully.

1. One star, our Sun, orbited by planets, moons, asteroids, comets, Kuiper Belt objects, and interplanetary gas and dust.   (Section 8.1)

2. Mass. A planet has enough mass to clear the region around its orbit, and a dwarf planet does not.   (Section 8.1)

3. All of the planets revolve around the Sun in the same direction. The orbits of all the planets lie nearly in the ecliptic plane.   (Section 8.2)

4. (a) Full Moon;   (b) first quarter.   (Section 8.4)

5. (a) 5;   (b) 3;   (c) 2;   (d) 1;   (e) 4;   (f) 6.   (Sections 8.6 through 8.9)

6. A combination of their forward motion and their motion in toward the Sun under the Sun's gravity.   (Section 8.9)

7. (a) 3;   (b) 1;   (c) 2;   (d) 4;   (e) 4;   (f) 2.   (Sections 8.8, 8.9)

8. Two days. While the Moon revolves around Earth, both Earth and the Moon revolve together around the Sun.   (Sections 8.10, 8.11)

9. Gravity.   (Section 8.12)

10. (a) 2;   (b) 1;   (c) 2;   (d) 2;   (e) 1;   (f) 2;   (g) 2.   (Sections 8.13 through 8.15; Table 8.2)

11. (a) 1;   (b) 6;   (c) 2;   (d) 4;   (e) 5;   (f) 3.   (Sections 8.2, 8.13, 8.14; Table 8.2)

12. (1) Irregular, rocky bodies that orbit the Sun, mostly between the orbits of Mars and Jupiter. (2) Small icy objects that orbit the Sun in a huge ring-shaped region from 30 AU to perhaps 50 AU near the ecliptic plane. (Sections 8.1, 8.15)

9

THE PLANETS

The Earth is the cradle of mankind, but one does not live in the cradle forever.

Konstantin Tsiolkovsky (1857–1935)

Objectives

☆ Compare and contrast the general properties and surface conditions of Mercury, Venus, Earth, and Mars.

☆ Explain what is meant by "morning star" and "evening star."

☆ Compare and contrast the atmospheres of Mercury, Venus, Earth, Mars, Jupiter, Saturn, Uranus, and Neptune.

☆ Describe conditions on Mars at spacecraft landing sites.

☆ Give two observations that indicate water might once have flowed on Mars.

☆ Compare and contrast the internal structure of Earth and Jupiter.

☆ Explain the theory of plate tectonics (continental drift) in relation to Earth's geological activity.

☆ Outline two environmental concerns related to Earth's atmosphere.

☆ List and give the current explanation for a famous feature visible in a small telescope for Venus, Mars, Jupiter, and Saturn.

☆ Compare and contrast the general properties of Jupiter, Saturn, Uranus, and Neptune.

☆ Tell what is known about the satellites of Mars, Jupiter, Saturn, Uranus, and Neptune.

☆ Describe known properties of Pluto and Eris.

9.1 PLANET MERCURY

Mercury, the planet closest to the Sun, is often hidden in the Sun's glare (Figure 9.1). Mercury was appropriately named for the swift Roman messenger god. It wings around the Sun fastest of all the planets, at a mean speed of 172,000 km/hour (107,000 miles per hour).

Our first close-ups of the whole planet are from U.S. orbiter MESSENGER, Mercury Surface Space Environment Geochemistry and Ranging Mission (2011–) (Figure 9.1).

Mercury looks like our Moon. The surface is ancient and heavily cratered. The craters suggest that meteorites bombarded the inner planets in the final stages of their formation. The largest crater, Caloris Basin, is 1300 km (800

Figure 9.1. Mosaics that cover all of Mercury's surface under daylight conditions, taken by MESSENGER. Craters, smooth dark plains, and large scarps are visible.

Figure 9.2. Impact crater John Lennon, named by the International Astronomical Union for the British pop music superstar. Mercury features are now named for noteworthy deceased artists, musicians, painters, and authors who have been recognized for over 50 years.

miles) across. Large, smooth areas resembling the Moon's maria suggest that extensive lava flooding occurred in the past (Figure 9.2).

Huge **scarps**, or cliffs, up to 2 km (1.2 miles) high and 1500 km (930 miles) long crisscross the planet. They apparently formed when Mercury's interior cooled and shrank, compressing the crust.

Mercury's axis of rotation is vertical (not tilted as Earth's is), so the Sun always shines exactly overhead at its equator and at a very low angle at its poles. The planet has no seasons.

Temperatures vary from extremely hot in direct sunlight, 700 K (430° C or 800° F), to bitter cold, 90 K (–180° C or –300° F), in the dark.

Solar wind blasts, micrometeoroid impacts, and high temperatures free volatiles from rocks, which creates a very thin, unstable atmosphere. Helium, sodium, hydrogen, oxygen, potassium, calcium, and magnesium have been detected. The surface air pressure is barely a trillionth of Earth's. A global magnetic field about 1 percent as strong as Earth's affects charged particles in the solar wind and the thin atmosphere's variable constituent abundances.

MESSENGER found (1) solid evidence of water ice deposits above and below the ground in permanently shadowed craters near the poles where the temperature of 125K (–148° C or –235° F) is cold enough to retain ice, (2) unfamiliar chemistry—a lot of the volatiles sodium, potassium, and sulfur in surface rocks, and (3) a strange landform with hollows or depressions in older-looking craters. They are 18 m to over 1.6 km (60 feet to over a mile) across and 18 m to 37 m (60 to 120 feet) deep, often with bright interiors and halos. ▶http://messenger.jhuapl.edu◀

Mercury has a crust of light silicate rock (average density about three times that of water). How can Mercury's average density be 5.4 times that of water?___

_____

_____

Answer: Mercury's interior must be very dense. Scientists figure that Mercury has a large, partly molten iron core (density eight times water's) that fills about 80 percent of the planet.

✪   Help astronomers map the history of Mercury. ▶http://cosmoquest.org/projects/mercury_mappers/◀

9.2 VENUS: OBSERVING

Brilliant **Venus** was named for the Roman goddess of love and beauty. At night Venus outshines all the stars. The planet is so conspicuous that it is often mistakenly reported as an unidentified flying object (UFO) (Figures 9.3 and 9.4).

Venus, like Mercury, circles the Sun inside Earth's orbit. As a result, both planets appear close to the Sun in our sky. They shine in the western sky just

Figure 9.3. Venus as an evening star in the western sky right after sunset.

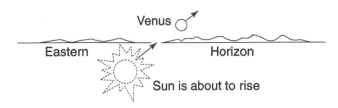

Figure 9.4. Venus as a morning star in the eastern sky just before sunrise.

after sunset near their eastern elongation. Then they appear to follow the Sun across the sky. They are frequently called **evening stars** at that time.

They are **morning stars** in the eastern sky just before sunrise near their western elongation. Then they lead the Sun across the sky.

Both Venus and Mercury go through a cycle of phases (Figure 8.7) that you can observe in a small telescope. Venus rotates from east to west, or **retrograde**.

Normally Venus and Mercury pass above or below the Sun at conjunctions. Mercury—about 13 times in a century—and Venus—twice in 122 years—transit, or pass directly in front of the Sun, at inferior conjunction. Observers see a tiny dot moving across the bright face of the Sun. Mercury will transit next on May 9, 2016, and Venus on December 11, 2117. ►http://eclipse.gsfc.nasa.gov/transit/transit.html◄

Venus is at inferior conjunction at intervals of 584 days. Then it comes closer to Earth, about 40 million km (25 million miles) away, than any other planet.

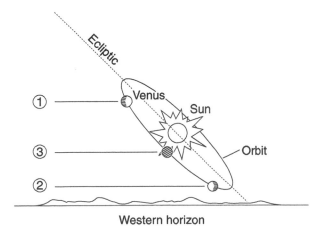

Figure 9.5. Venus in orbit.

(a)   (b)   (c)

Figure 9.6. Venus as viewed by (a) Earth-based telescope; (b) U.S. Pioneer Venus Orbiter; (c) computer-generated map from Magellan radar imagery and altimetry data.

Refer to Figure 9.5. Determine the location of Venus when it is (a) an evening star _____ ; (b) a morning star _____ ; (c) at conjunction _____.

Answer: (a) 1; (b) 2; (c) 3.

9.3 VENUS: THE PLANET

Venus shines brilliantly because it is shrouded in thick clouds that reflect a lot of sunlight. These perpetual clouds hide the surface from our view.

More than 20 U.S. and Russian robot spacecraft have successfully encountered Venus and transmitted data back to Earth for analysis and image processing. Spacecraft firsts at Venus include Mariner 2 / flyby 1962, Venera 3 / impact 1965, Venera 7 / soft landing 1970, Venera 9 / panorama 1975, and Venera 13 / color images of surface 1981 (Figure 9.6).

ESA robot orbiter Venus Express (2004–) confirmed that the atmosphere is about 97 percent carbon dioxide and 1 to 3 percent nitrogen with traces of water vapor, helium, neon, argon, sulfur compounds, and oxygen. It circulates

Figure 9.7. The first look at the surface of Venus, from the short-lived Venera 9 spacecraft.

in large global motions. Cloud layers about 19 km (12 miles) thick (altogether) are about 50–70 km (40 miles) above the surface. Apparently they are colored yellow by corrosive sulfuric acid.

Venera landers first took surface pictures in 1975. Rocks and soil look orange beneath the thick clouds. In direct sunlight on Earth, they would look gray. The landers found a very inhospitable world. They all expired within two hours because of hellish conditions (Figure 9.7).

Surface temperatures reach 755 K (482° C or 900° F) because carbon dioxide and water vapor in the clouds let in visible light from the Sun but do not let out infrared heat that is given off by the hot rocky surface. This **greenhouse effect** makes Venus ever hotter. Atmospheric pressure is a crushing 90 times more than Earth's (over 90 atmospheres, or 1330 pounds per square inch). There are many lightning bolts and thunderclaps.

To map the surface, scientists bounce radar signals from Earth or robot spacecraft off Venus and analyze the echoes. Our best radar imaging data

Figure 9.8. Arachnoids, found only on Venus. A web of fractures surrounds spiderweb-like concentric ovals 50 km to 230 km in diameter, on radar-dark plains on Venus.

Figure 9.9. Western Eistla Regio scene developed from Magellan radar imagery and altimetry data. A rift valley in the foreground extends to the base of Gula Mons, a 3-km-high volcano at the right. Sif Mons, a 300-km-wide and 2-km-high volcano, is at the left.

came from the U.S. Venus orbiter Magellan (1989–1994), which mapped 99 percent of the planet (Figures 9.8 and 9.9) and also studied the interior and atmosphere. ▶www2.jpl.nasa.gov/magellan◀

Venus was completely resurfaced a few hundred million years ago. Radar images show that all the terrain is dry and rocky. About 80 percent is lava plains with long, winding channels, impact craters, and volcanoes. The difference between the lowest and highest elevations is 15 km (9 miles).

Surface features on solar system bodies are customarily given two names by the International Astronomical Union. One is descriptive and is used for geographic features on all bodies. The other name is for identification. On Venus names of surface features honor love goddesses or famous women who have been dead for at least three years. The sole exception is Maxwell Montes, the largest mountain range, which honors the Scottish physicist James Clerk Maxwell (1831–1879), whose theories of electromagnetism made radar possible.

Impact craters on Venus look different from those on other worlds. Small chunks of rock from space burn up in the thick atmosphere. Large impactors crash onto Venus at about 20 km (12 miles) per second, releasing a tremendous amount of energy that vaporizes the crashing object and the surrounding ground. **Ejecta**, surface material, is blasted out. It stays molten because of the very high surface temperature and flows away from the crater in patterns that look like flower petals.

Highlands that appear like continents tower above the dry plains. Terra Aphrodite, the largest, is half the size of Africa. Smaller Ishtar Terra is the size of the continental United States. Here Maxwell Montes, a mountain massif, soars nearly 11 km (7 miles) above the mean radius. There are many fault zones.

Numerous strange volcanic features dot the plains. **Shield volcanoes**, small domes 2–3 km (1–2 miles) across and a few hundred meters high, are caused by a thin, runny lava that builds a broad mountain. **Pancake domes**, typically 25 km (15 miles) wide and less than a mile high, are associated with a thick, sticky magma with more than the normal silica content. **Coronae**, circular rings of low-relief folds, were most likely formed when lava flows created a dome, which then sank and collapsed. Some volcanoes may still be active.

Venus is close to Earth in size, mass, density, and distance from the Sun. However, you could not live there comfortably. Give three reasons why not.

(1) _____ ;

(2) _____ ;

(3) _____

Answer: Venus (1) is much too hot: 482° C (900° F); (2) has a poisonous carbon dioxide atmosphere; and (3) has a crushing atmospheric pressure, over 90 times Earth's.

9.4 PLANET EARTH

Our own planet **Earth** shines like a rare blue and white jewel in space (Figure 9.10). Third from the Sun, it is the most important planet of all to us.

The total surface area of our planet is almost 5.10×10^8 km² (199 million square miles). More than 70 percent of our planet is covered by water, which is unique in the solar system.

The highest mountain on Earth is Mt. Everest in Asia, almost 9 km (29,000 feet) above sea level. The deepest measured underwater spot is the Marianas Trench, more than 11 km (36,000 feet) below the Pacific Ocean's surface.

Earth's mass is about 6×10^{24} kg. This mass provides the surface gravity we are used to.

Figure 9.10. Earth viewed from the Moon by Apollo astronauts.

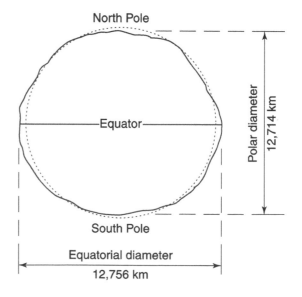

Figure 9.11. Earth's daily rotation around its axis has produced an equatorial bulge and polar flattening (exaggerated).

Refer to Figure 9.11. How much longer in kilometers is the distance across Earth's equator than the distance from its North to South Pole? _____

Answer: About 43 km.

Solution:

Equatorial diameter – Polar diameter = 12,756 km – 12,714 km = 42 km.

9.5 EARTH'S STRUCTURE

Astronomers figure that Earth was born about 4.6 billion years ago. It formed together with the other planets out of the same contracting cloud of gas and dust that formed the Sun (Section 4.3).

Geologists cannot go inside Earth to examine it directly. Instead, they determine its structure and composition from the way seismic waves, from earthquakes and explosions, are transmitted through the Earth and along its surface. They picture Earth today in three main layers: The **crust** is the thin, outermost, solid layer. It is an average of 35 km (22 miles) thick, being thicker where there are continents and thinner under the oceans. The crust is composed mainly of lightweight rocks such as granite and basalt. The **mantle** is the layer below the crust. It extends in about 2880 km (1800 miles). Laboratory analysis of samples from volcanoes indicates that the thick mantle consists mostly of dense silicate rock that behaves somewhat like taffy—yielding under steady pressure but frac-

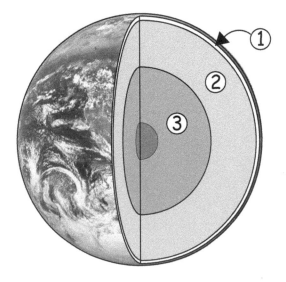

Figure 9.12. The structure of Earth showing three principal layers.

turing under impact. The central layer, which is 3470 km (2170 miles) thick, is called the **core**. Here an outer, molten, metallic layer about 2100 km (1300 miles) thick probably surrounds a solid center. The core is probably made of dense iron and nickel at a temperature of about 6400 K.

Refer to Figure 9.12. Identify the three principal layers of Earth and state the approximate thickness of each. (1) _____ ; (2) _____ ; (3) _____

Answer: (1) Crust—an average of 35 km (22 miles); (2) mantle—about 2880 km (1800 miles); (3) core—about 3470 km (2170 miles).

9.6 EARTH'S GEOLOGICAL ACTIVITY

The surface of our restless Earth is constantly changing because of erosion and geological activity. The oldest rocks discovered so far amid the remote lakes and tundras of northwest Canada are about 3.96 billion years old.

Substantial evidence indicates that about 200 million years ago all of the world's continents were joined in one huge supercontinent called Pangaea, which later broke up.

According to the theory of **plate tectonics**, also called the **continental drift theory**, continents and ocean floor are embedded in plates, or rock slabs, several thousand miles across. The plates move slowly on the slightly yielding mantle beneath. Earth's crust is reshaped at plate boundaries. Where the plates move apart, the continents separate slowly, at about 2.5 cm (1 inch) per year. That adds up to over 5000 km (3000 miles) in 200 million years.

Movement of the plates is also responsible for mountain building, earthquakes, and volcanic activity. These events occur at boundaries between the moving plates where they press on each other forcibly.

A popular theory says that magma convection currents power the continental drift. Magma currents flow up through the mantle. Upon meeting cool, rigid rocks, they flow horizontally. Friction drags the continent-bearing plates along. Finally the cooled magma sinks. Along midocean ridges, magma pours through the crust continually creating new rocks.

Remarkable evidence confirms that the continents drift. Laser-ranging devices bounce light pulses off reflectors on the Moon and measure drift. The coastlines of South America and West Africa seem to fit together, and similar plant and animal fossils are found along both, although they are now separated by almost 5000 km (3000 miles) of Atlantic Ocean (Figure 9.13).

The ages of rocks from the bottom of the Atlantic Ocean have been measured. The oldest, found near the continent coastlines, are 150 million years old.

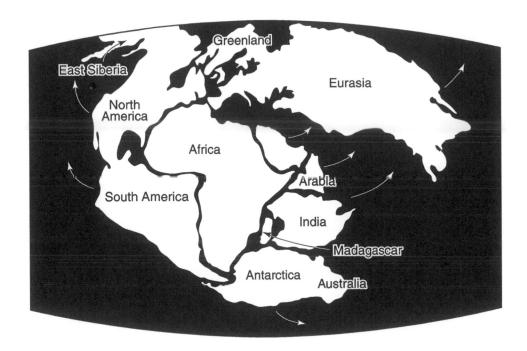

Figure 9.13. A map of Earth as it probably appeared about 200 million years ago.

If 4-billion-year-old rocks had been found on the bottom of the Atlantic Ocean, how would the continental drift theory have been affected? Explain.

_____

_____

_____

_____

Answer: Serious doubts would have been raised about the correctness of the theory, which says that the Atlantic Ocean, almost 5000 km (3000 miles) wide between the continental coastlines, formed in the last 200 million years. It did not exist 4 billion years ago.

9.7 EARTH'S MAGNETISM

Our planet has a **magnetic field**, or region of magnetic forces, that affects compass needles.

The magnetic poles are located at about 83°N latitude, 115°W longitude in northeast Canada—1300 km (800 miles) from the geographic North Pole—and at about 65°S, 138°E in Antarctica. The magnetic poles drift some 40 km (25 miles) a year. Earth's magnetic field is believed to be generated by its liquid iron-nickel core, which acts like a giant dynamo as the planet spins. The complex motion of this core probably causes the long-term migration of the magnetic poles.

Earth's **magnetosphere**, the region around the planet where its magnetic field is influential, extends out into space about 4 Earth radii on the sunward side. The **magnetotail**, the part of the magnetosphere on the side away from the Sun, extends like a tail from 10 to perhaps 1000 Earth radii.

Many energetic, charged particles from the solar wind that could be deadly are trapped by Earth's magnetic field. They keep moving around rapidly inside two doughnut-shaped regions called the **Van Allen belts** in the magnetosphere.

What is the magnetosphere? _____

_____

Answer: The region surrounding Earth where the magnetic field is influential.

9.8 EARTH'S ATMOSPHERE

Earth is surrounded by an **atmosphere** that extends several hundred miles out into space.

Earth's first atmosphere over 4 billion years ago was probably very different from air today. Noxious compounds of hydrogen, carbon, oxygen, and nitrogen—such as carbon dioxide, ammonia, and methane, plus water vapor—may have **outgassed**, or been released from the interior of the hot, young planet.

The carbon dioxide, which is highly soluble in water, could have been removed by combining with substances such as calcium in the ocean to form limestone. The free oxygen we need for respiration probably was generated by green plants. In photosynthesis, green plants absorb carbon dioxide from the air, utilize it for growth, and release oxygen.

Air today contains about 78 percent nitrogen, 21 percent oxygen, and 1 percent argon, carbon dioxide, and other gases. It also has variable amounts of water vapor, dust, carbon monoxide, chemical products of industry, and microorganisms.

Over half of this air is packed within the first 6 km (4 miles) above Earth's surface. The air thins out fast with increasing altitude. At about 12–50 km (7–30 miles) above sea level, the Sun's ultraviolet light acts on air to produce **ozone**, a molecule made up of three atoms of oxygen. A global ozone layer shields people and plants from harmful ultraviolet radiation from the Sun. Beyond 160 km (100 miles), satellites can orbit without being dragged down.

Researchers are using sophisticated computer simulations and instruments on the ground and aboard airplanes and spacecraft to study potentially dangerous changes in the atmosphere and climate caused by human activity. Elevated levels of ozone-destroying chlorofluorocarbons released by refrigerators, air conditioners, and some aerosol sprays may cause the observed gaping holes in the ozone layer over the polar regions and thinning over the midlatitudes. Increasing concentrations of carbon dioxide and impurities released by burning coal and oil and the clearing away of rain forests might cause a global warming of our planet similar to the greenhouse effect on Venus.

The total mass of the entire atmosphere is about 5000 trillion tons. Gravity keeps the atmosphere tied to Earth, although atoms occasionally escape at the top. At sea level all that air presses down with a force of 1.03 kg/cm^2 (14.7 pounds per square inch), called 1 atmosphere of pressure. The millibar is another common unit of atmospheric pressure. At sea level, the air pressure on Earth is about 1013 millibars.

What is the (a) composition and (b) pressure at sea level of the air that supports our lives on Earth? (a) _____

_____

(b) _____

_____

Answer: (a) About 78 percent nitrogen, 21 percent oxygen, and 1 percent carbon dioxide and other gases; variable amounts of water vapor and impurities. (b) About 1.03 kg/cm^2 (14.7 pounds per square inch), also called 1 atmosphere and 1013 millibars.

9.9 MARS: OBSERVING

Red **Mas** reminded the Romans of blood and fire, so they named the planet after their god of war. It has two small moons, appropriately called Phobos ("fear") and Deimos ("terror"), which can be seen only in powerful telescopes.

Superior planets like Mars look brightest when they are on the opposite side of Earth from the Sun, a position called **ops t on**. Then a fully lighted disk faces us. Mars is at opposition at intervals of 780 days on the average.

Superior planets are hardest to observe when they are on the opposite side of the Sun from Earth, a position called conjunction (Figure 9.14).

Figure 9.14. Two important aspects of Mars from Earth.

Mars comes closer to Earth at some oppositions than it does at others because of the eccentricity of its orbit (Table 8.2). Close oppositions are called **favorable**, because the disk of Mars looks larger, and observing is better. The most favorable oppositions occur when Mars is near perihelion (Figure 9.15). Then Mars is only about 56 million km (35 million miles) away from Earth. This happens toward late summer at intervals of 15 to 17 years.

When Mars is near **favorable opposition** you can, with a telescope, see features that have long excited the imagination. In each hemisphere, white polar caps grow in the Martian winter and shrink in summer. Dust storms and dark areas once mistakenly thought to be flowing water or vegetation vary. The dark areas are mountains and rocks exposed after dust storms (Figure 9.16).

"Canali," straight, dark channels, were first reported in 1877 by Italian astronomer G. V. Schiaparelli. This word was mistranslated into English as "canals." U.S. astronomer Percival Lowell (1855–1916) caused great excitement at the beginning of the twentieth century when he mistakenly said that intelligent Martians had built the canals. More U.S. and Russian probes have

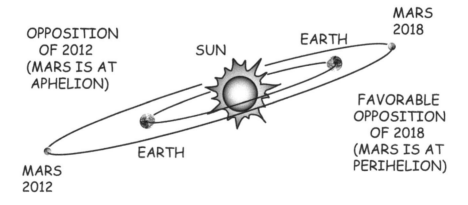

Figure 9.15. Favorable and unfavorable oppositions of Mars in the current cycle.

Figure 9.16. Mars at six oppositions. Mars was closest, with largest apparent angular size, in 2003, when it came within 56 million km (35 million miles) of Earth.

explored Mars than any other planet. None found canals or Martians. Still Mars exploration inspires a new generation.

Why is Mars best observed at favorable oppositions? _____

_____

Answer: It is closest to Earth then. (Both Mars and Earth travel in elliptical orbits around the Sun, so the distance between them varies considerably.)

9.10 MARS: THE SURFACE

We took our first good look for life on the surface through the "eyes" of the U.S. robot Viking Lander 1, which set down on Mars on July 20, 1976.

Scattered rocks, powdery dirt, sand dunes, and distant low hills came into view on Chryse Planitia (Plains of Gold), a floodplain at 22.46N latitude,

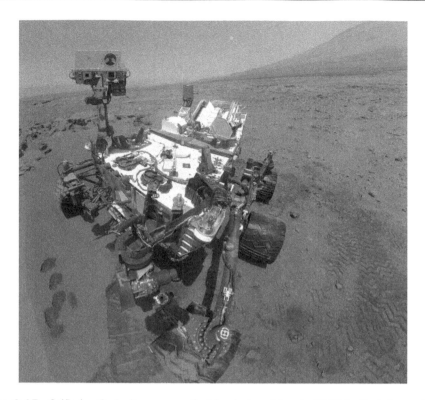

Figure 9.17. Selfie by Curiosity, a rover that landed on Mars in 2012 to find signs of habitability. Curiosity scooped up soil and drilled rocks for its onboard laboratory to analyze.

48.01W longitude. Two months later Viking 2 landed 7500 km (4600 miles) northwest near rocks pitted by gaseous volcanoes or meteorite impacts.

Viking 1 and 2 sent back over 4500 pictures, 3 million weather reports, and data from chemical and biological tests. The historic Viking Lander 1 became the first museum exhibit located on another world in 1984 when its ownership was transferred to the U.S. National Air and Space Museum.

Since 1997 four U.S. robot geologists have gone to Mars to look again. Outfitted with cutting-edge scientific gear, they wheeled around equatorial landing sites on opposite sides of Mars (Figure 9.17). They found evidence of past flowing water in rocks, minerals, and geologic landforms.

Panoramic views of the landing sites show rusty red soil that looks like iron-rich clay. Fine-grained reddish material covers rocks. Chemical weathering of the rocks and erosion seem to have occurred. Rounded pebbles, abundant sand, and dust particles argue for a previously water-rich planet. The fine soil is about 45 percent silicon oxide and 19 percent hydrated iron oxide (rust). The sky is colored pink in daytime by red dust that hangs in the atmosphere like smog. Sunsets are pale blue.

Summer temperatures at noon reach a maximum of 30° C (86° F). At night temperatures can drop to –123° C (–190° F) and a fine layer of frost appears. Air pressure is only about 7 or 8 millibars.

In 2008 U.S. Phoenix Mars landed in the intriguing arctic region. Its robotic arm dug into the soil and delivered samples to scientific instruments for analysis. Phoenix found water ice and chemicals that are important for life, plus minerals that are created in liquid water. Mars may have been or may still be a possible habitat for life.

No signs of large organisms were apparent at the sites. There was no flowing water, but there might have been in Mars's ancient past. It might be preserved underground today.

Briefly describe the surface of Mars at landing sites. _____

_____

_____

Answer: It looks like a red, dry, rock-strewn desert. The sky is pink. The temperature is cold.

9.11 MARS: THE PLANET

High-resolution images and 3-D maps from measurements by 20 robot orbiters at Mars show a harsh, rugged, dry planet. ►http://mars.jpl.nasa.gov◄

Although the southern hemisphere is mostly cratered and ancient, the northern hemisphere is basically smooth plains and typically several kilometers lower. The northern lowlands may be an ancient ocean basin filled in by sedimentation.

Mars has huge volcanoes. Olympus Mons (Mount Olympus) is the largest volcano in the solar system. It towers almost 25 km (16 miles) above the mean surface and contains more lava than the U.S. Hawaiian Islands.

Gravity data, derived from changes in Mars Global Surveyor's orbits, and extremely tall, massive volcanoes indicate that the planet's crust is about 50 km (30 miles) thick and does not drift as Earth's continents do. Mars probably has a mantle that is cooler and thicker than Earth's.

Linear bands of highly magnetized material appear in some of the oldest crust. In the past, Mars must have had a molten iron core and magnetic and geologic activity, although none is evident now.

The planet has deep canyons. The largest, Valles Mariner (Mariner Valley), is a complex network of rocky valleys extending 5000 km (3000 miles) around the equator with an average depth of 6 km (4 miles). Diverse areas evidence great landslides, faulting, and flow channels carved deeper by erosion.

Craters suggest that meteorites bombarded the planet. The lowest point

Figure 9.18. Deep, winding channels and craters on Mars.

on Mars, 7 km (4 miles) below the mean surface, is the bottom of the circular Hellas Planitia basin. Hellas, the largest impact crater in the solar system, is about 4 billion years old. Younger craters, such as the 18-km (11-mile) wide Crater Yuty, look as if water and shattered rock poured out and flowed for great distances after a big impact.

Mars does not have liquid surface water today. However, there is much evidence of ancient catastrophic flooding. Its deep, winding channels and gullies resemble river beds with tributaries (Figure 9.18). They look as if they were carved long ago by floodwaters and streams. Some rock types form only underwater. That water may be locked in the ice caps and permafrost under the surface. Climate changes may have turned a running water environment into the cold, dry world Mars is today.

Water is evident in solid and vapor form. The permanent ice caps at the north and south poles are made of frozen water. They are covered in winter by layers of carbon dioxide that freeze out of the atmosphere. Winter frost is apparently frozen water and dust. There are also occasional fog and filmy clouds.

The atmosphere is too thin to block the deadly ultraviolet rays from the Sun that beat down on the planet. It is made up of about 95 percent carbon dioxide, with 2 to 3 percent nitrogen, 1 to 2 percent argon, and 0.1 to 0.4 percent oxygen, with traces of water vapor and other gases.

Wild dust storms swirl out of the southern hemisphere in the summer. Often they rage over the whole planet. Winds up to 120 km (75 miles) per hour blow light-colored dust about, sculpting and exposing dark rock. Thin layers of ice and dust hundreds of kilometers long have been laid down at the north and south poles by global dust storms in alternating seasons.

A piloted flight to Mars could realistically take place in the next decade. First, new robots must go to send us more mapping and surface data. A robot could bring Martian rocks and soil back to Earth for analysis.

 Connect to Mars exploration: ▶http://mars.jpl.nasa.gov/participate◀

Scientists think water is critical for life. Perhaps life formed on Mars in the distant past when that planet was warmer and wetter. Possibly microbes still survive.

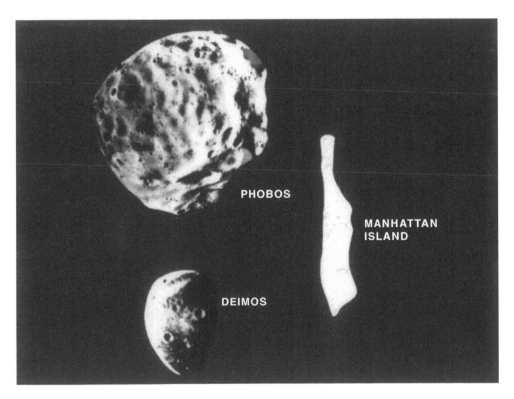

Figure 9.19. Phobos and Deimos, the moons of Mars, shown at the same scale as the island of Manhattan.

List two pieces of evidence that indicate water once flowed on Mars.

(1) _____

_____

(2) _____

_____

Answer: (1) The deep, winding surface channels look as if they were carved by great rivers. (2) The permanent polar ice cap at the north pole is made of frozen water that may once have flowed on Mars.

9.12 MOONS OF MARS

Phobos and Deimos are small, irregular rock chunks only about 28 km (17 miles) and 16 km (10 miles) long, respectively (Figure 9.19). Phobos orbits Mars every 7.7 hours, while Deimos completes a circuit in 1.3 days.

Both moons look fairly old, with many impact craters of varying ages. Phobos has striations and chains of small craters. Stickney, its largest crater, measures practically 10 km (6 miles) across.

Two moons of Mars were mentioned by English writer Jonathan Swift in 1727, long before they were actually discovered in 1877 by U.S. astronomer Asaph Hall (1829–1907).

Briefly describe the moons of Mars. _____

_____

Answer: Small, irregular, cratered rock chunks.

9.13 JUPITER: OBSERVING

Giant **Jupiter** was named for the mythological Roman king of the gods and ruler of the universe. The biggest planet of all, Jupiter at night outshines the stars and all the planets except Venus.

Jupiter's colorful, parallel, dark and light cloud bands, Great Red Spot, and four biggest moons can be seen through a small telescope. Io, Europa, Ganymede, and Callisto change patterns nightly as they circle the planet. Astronomical publications and computer software (see Useful Resources) list current moon positions, occultations, and transits.

Seven U.S. robots flew by within a few hundred thousand km (miles) of Jupiter's cloud tops (1973–2007). All sent back superb global views and voluminous data on the chemistry, atmosphere, magnetosphere, and interior of Jupiter and its moons.

A probe plunged into the clouds and transmitted data for an hour until heat and pressure killed it. An orbiter sent data and images of Jupiter and its moons for eight years. ▶http://solarsystem.nasa.gov/galileo◀

Juno (2011–) is to study Jupiter from a unique polar orbit in 2016, going to within 5000 km (3100 miles) of the cloud tops.

Through a small telescope Jupiter's four brightest moons look like stars. What observations show they are really satellites of the planet?

Answer: The moons are observed to change positions nightly as they circle the planet.

Figure 9.20. Visible light image of storms on Jupiter from Hubble Space Telescope.

9.14 JUPITER: THE PLANET

Jupiter is more massive than all the other planets and their moons combined. It just missed being a star. If Jupiter were about 80 times more massive, nuclear fusion reactions could have started.

The planet's thick atmosphere is made mostly of hydrogen and helium. Below a liquid encloses an Earth-size solid core. A thin ring system comprised of dust grains that meteoroids blast off of the inner moons encircles Jupiter. The outermost part, a gossamer ring just beyond a brighter one, extends to some 210,000 km (130,000 miles) from the planet's center.

Colorful, changing cloud features and convoluted weather patterns circulate in the dynamic, observable atmosphere. Superbolts of lightning flash. Complex patterns show in and between the moving dark-colored **belts** and **lighter zones**. Hydrogen, helium, the detected traces of methane, and water vapor are all colorless. Sulfur or phosphorous compounds and ammonia at various depths must give the atmosphere its bright red, orange, yellow, and brown colors and white clouds. The famous **Great Red Spot** is a colossal atmospheric storm. It has been observed for over 300 years at varying sizes, brightness, and color. The Great Red Spot rotates counterclockwise and also goes around with the planet. Cooler than surrounding clouds, it towers up to 24 km (15 miles) above them. Smaller storms and eddies appear throughout the banded clouds (Figure 9.20).

Temperatures hit 160 K (–170° F) at the cloud tops. The atmosphere extends down about 21,000 km (13,000 miles). The density of hydrogen increases steadily from the top inward as the pressure increases, until it changes to liquid hydrogen. The pressure below must be high enough to compress hydrogen to an extraordinarily dense form called liquid metallic hydrogen.

At the core temperatures may be 30,000 K (53,000° F), which would explain the observation that Jupiter radiates about twice as much heat as it receives from the Sun. The planet has a powerful magnetic field that traps ions and electrons in a complex system of large, intense radiation belts. Plasma (a collection of ions and electrons) oscillations account for some of Jupiter's observed radio emission. The magnetic field is essentially dipolar but opposite Earth's in direction. Electrical currents in the liquid hydrogen layer could be its source. At the cloud tops, Jupiter's magnetic field is 1.5 to 7 times more powerful than Earth's. Jupiter's enormous magnetosphere varies in size, possibly due to changes in the solar wind pressure. It may stretch sunward 7 million km (4 million miles) and outward nearly 650 million km (400 million miles) to Saturn's orbit.

Jupiter's atmosphere is especially interesting, because it may be similar to Earth's primitive one.

What is Jupiter's atmosphere made of?_____

Answer: Mostly hydrogen and helium, with traces of methane, ammonia, water vapor, and other gases.

Figure 9.21. Comparison portrait of Jupiter's four Galilean moons: Io, Europa, Ganymede, and Callisto. (Jupiter is not at the same scale as the satellites.)

9.15 JUPITER'S MOONS

There are at least 63 confirmed moons orbiting Jupiter (Table 8.3). Most are small. The Galileo orbiter focused on the four biggest, collectively called the **Galilean moons** after discoverer Galileo Galilei, plus innermost Metis, Amalthea, and Thebe (Figure 9.21).

Small Amalthea resembles a dark red football with meteorite impacts.

Colorful Io has active volcanoes that spew sulfur-rich materials that color the surface bright orange, red, brown, black, and white. Io's bright white spots are sulphur dioxide frost, and its tenuous atmosphere is primarily sulphur dioxide gas. The volcanoes may be due to the heating that occurs as Europa and Ganymede tug gravitationally on Io and Jupiter alternately pulls Io back to its regular orbit. This pumping creates tidal bulges on Io's surface that are up to a hundred times greater than the typical 1 m (3.3 feet) tidal bulges on Earth.

A gigantic cloud of charged particles, mostly ions of sulphur and oxygen, wobbles around Jupiter at Io's distance. The particles are likely stripped off Io by magnetic forces as Jupiter's magnetosphere rotates with the planet. Cloud particles may also travel along Jupiter's magnetic field lines into its north and south polar atmospheres, causing brilliant Jovian auroras.

There is evidence of water ice on the surfaces of Europa, Ganymede, and Callisto. Europa, about the same size and density as our Moon, is the brightest Galilean moon. Its smooth, icy crust, crisscrossed by long lines, may hide a global ocean of water warmed by tidal heat.

Ganymede and Callisto most likely have a rocky core with a water/ice mantle and a crust of rock and ice. Ganymede is the largest known moon in the solar system, 5262 km (3262 miles) in diameter. It has dark, probably ancient, areas with many craters and lighter, younger terrain that is grooved, suggesting global tectonic activity. Callisto's surface looks oldest, with numerous impact craters. The largest craters may have been erased by the flow of icy crust. Features that look like the remains of very large basins may record collisions with large chunks of rock and metal.

(a) What is the largest moon in the solar system? _____

(b) What is its diameter? _____

Answer: (a) Ganymede; (b) 5262 km (3262 miles).

9.16 PLANET SATURN

Saturn, the most distant bright planet, was named for the Roman god of agriculture. Dazzling rings surround Saturn (Figure 9.22). The brightest were named in order of their discovery. From the planet outward, they are known as D, C, B, A, F, G, and E.

We see the rings at various angles, from edge-on to 29°, as both Earth and Saturn orbit the Sun. The A and B rings separated by a gap called the Cassini Division and the fainter C ring are visible in small telescopes. Since Saturn takes 29.5 years to orbit the Sun, we see it in essentially the same orientation toward Earth in the same area of our sky for months. Although the brightest rings span 282,000 km (175,000 miles) wide, they are only about 10 m (33 feet) thick. Stars can be seen through them (Figure 9.22).

Saturn's rings consist of billions of dust- to house-size water ice particles that resemble icy snowballs or ice-frosted rocks, orbiting Saturn. They shine by reflecting sunlight. Probably the larger particles are the remnants of moons,

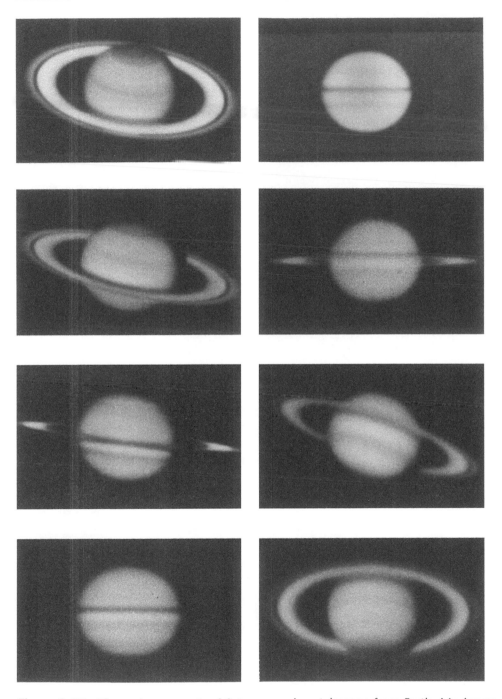

Figure 9.22. The various aspects of Saturn seen in a telescope from Earth. Maximum inclination, 27°, of the south side of the rings toward the Sun occurred in 2003; edge-on in 2009; north side of the rings toward the Sun will occur next in 2017.

asteroids, and comets that were shattered by impacts, collisions, and tectonic activity. The rest may be material that never collected into a single moon.

Voyager 1 in 1981 and Voyager 2 in 1982 sent back 33,000 images of the Saturnian system as they flew by. The main rings contain thousands of tiny ringlets, which are partly intertwined and kinked by the gravitational forces of small moons that shepherd the ring material. Long, spokelike features in the B ring may be shiny, fine particles raised by electrostatic forces (Figure 9.23).

U.S./European robot Cassini (1997–), using gravity assists from Venus, Earth, and Jupiter, neared Saturn in 2004 and ejected a probe, Huygens. After 20 days, Huygens parachuted through the clouds of the moon Titan. Huygens gathered information about the chemical composition of the atmosphere and clouds and snapped pictures for two and a half hours as it descended. Upon landing on the frozen ground, Huygens sent a surface report for a few minutes. Cassini orbiter is still radioing back data on Saturn and its satellites. ▶http://saturn.jpl.nasa.gov◀

Like Jupiter, Saturn is a huge, multilayered gas ball of mostly hydrogen with a few percent of helium. A dynamic atmosphere, lattened at its poles by

Figure 9.23. Rings of Saturn from Voyager, computer enhanced to bring out fine details. The brighter rings are about five times wider than Earth (shown to the same scale), but their thickness is barely 1 km.

rapid rotation, has rain, lightning, and snow. A hazy layer above the visible clouds dims colors and features such as belts, zones, and long-lived ovals. High pressure and temperature below create a central iron-silicate core enclosed in metallic and liquid hydrogen layers. Saturn radiates more energy than it absorbs from the Sun, likely due to internal heat.

At about 29.5-year intervals, when Saturn's northern hemisphere receives the most heat from the Sun, a large white spot suddenly appears. Giant waves of gas rise up from deep in the atmosphere. The Great White Spot storm of 2010 spread thousands of kilometers across the planet. Temperatures spiked to a new record high. It faded in a few months.

Highest speed winds, over 1600 km (1000 miles) per hour, occur at the equator and are much stronger than Jupiter's. Auroral emissions occur near the poles when a blast of charged particles in the solar wind strike Saturn's magnetic field and atmosphere.

With a mass equal to 95 Earths in a volume 844 times Earth's, Saturn has the lowest average density of all the planets. It could float in water if a big enough sea existed anywhere.

Saturn's magnetosphere is about one third as big as Jupiter's. It, too, varies in size when the solar wind intensity changes. It may extend sunward nearly 2 million km (a million miles). The magnetic field drags along charged particles, which circle Saturn as the planet rotates.

(a) What are Saturn's rings made of? _____

_____

(b) Can you explain why they look solid in a small telescope?_____

_____

Answer: (a) Billions of dust- to house-size icy particles that resemble icy snowballs or frosted rocks orbiting Saturn. (b) The particles are so numerous and so far away from us. (Remember, distant galaxies look solid, too, although they are made of billions of separate stars.)

9.17 SATURN'S MOONS

Saturn has at least 60 confirmed and several suspected moons (Table 8.3). Others may be discovered as scientists continue analyzing the voluminous Voyager and Cassini encounter data.

Titan is biggest and intriguing (Figure 9.24). It has a thick, hazy, orange-colored atmosphere that is mostly nitrogen, with hydrocarbons such as methane. The cold moon, 94K (−292° F), is made of rock and ice. Cassini saw mountains, river channels, and methane/ethane lakes. Methane and ethane form clouds, rain

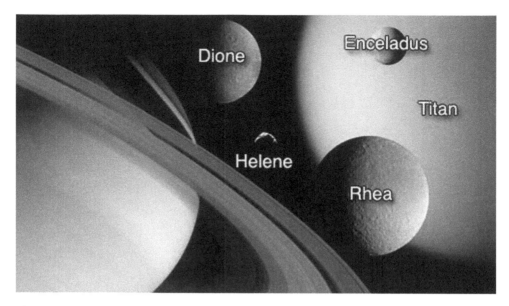

Figure 9.24. Saturn's largest moon, Titan, and some of its medium-sized moons, in this composite of Cassini images.

down, and evaporate into clouds again. Fields of dark hydrocarbon dunes cover vast areas. Data suggests an internal ocean and volcanic outgassing.

The other large moons are apparently mainly water ice. Except for Enceladus, all are heavily cratered. Geysers at a hot spot at the south pole of Enceladus shoot ice crystals far out, perhaps evidencing liquid water beneath its surface. Hyperion looks oldest, with evidence of meteoritic bombardment. Iapetus has icy and dark material on opposite sides.

Irregular shapes of small moons indicate they are fragments of shattered larger bodies. Prometheus shepherds the inner edge and Pandora, the outer edge of the F ring. Their gravitational effects at varying distances may cause the ring's kinks.

Which of Saturn's moons is the largest and very intriguing? Explain.

_____

_____

Answer: Titan is similar to the early Earth in many ways, except with liquid methane instead of water. Life here is based on liquid water. Prebiotic processes may be going on there based on liquid methane.

9.18 PLANET URANUS

Uranus was the first planet identified by means of a telescope. British astronomer William Herschel (1738–1822) discovered it in 1781 using a 150-mm (6-inch) telescope he made himself. Almost named for King George III, Uranus was finally named traditionally for the Greek god of the heavens.

Uranus, with a maximum magnitude of +5.7, looks like a small disk (sometimes tinted blue) through a telescope. You might spot it with your eye or binoculars if you know exactly where to look (see Useful Resources).

The planet was largely a mystery until Voyager 2 flew within 81,500 km (50,600 miles) of its cloud tops in 1986. Voyager 2 sent back 7000 images of the Uranian system.

Tipped on its side and surrounded by a system of narrow rings, Uranus resembles a giant bull's eye. The angle between its axis and the pole of its orbit is a unique 98°. The north and south polar regions are alternately exposed to sunlight and darkness as Uranus orbits the Sun. Its rotation is retrograde.

Possibly early in its history Uranus suffered a collision with a planet-sized body that knocked it over.

The atmosphere is mostly hydrogen and about 15 percent helium, with smaller amounts of methane and other hydrocarbons. It looks blue because methane preferentially absorbs red light from sunlight. The atmosphere has clouds running east to west like those of Jupiter and Saturn.

Winds blow in the same direction as the planet rotates, at speeds of 40 to 160 m/sec (90 to 360 miles per hour). Surprisingly, sunlit and dark cloud tops show the same average temperature, about 60 K (–350° F).

Voyager 2 detected haze around the sunlit south pole and large amounts of ultraviolet light, called **dayglow**, radiated from the sunlight hemisphere.

Uranus has a magnetosphere with intense radiation belts and radio emissions. Its magnetic field axis is tilted 60° to the rotational axis. The magnetic field is comparable to Earth's in intensity, but it varies much more because it is so off center. It may be generated by an electrically conductive, superpressurized ocean of water and ammonia located between the atmosphere and rocky core.

A rotating cylindrical magnetotail extends at least 10 million km (6 million miles) behind the planet. It is twisted into a long corkscrew shape by the planet's extraordinary rotation.

Uranus's narrow rings are distinctly different from Jupiter's and Saturn's (Figure 9.25). They are very dark and are composed mainly of large icy chunks several feet across. Intense irradiation may have darkened any methane trapped in their icy surfaces. Transient dusty lanes appear and disappear. The chunks must bump into one another and make the fine dust that seems to be spread throughout the ring system. Atmospheric drag due to a hydrogen corona that Voyager 2 observed around Uranus may cause dust particles to spiral in to the planet.

Incomplete rings and varying opacity in several of the main rings suggest that the ring system may have formed after Uranus. Ring particles may be remnants of a moon that was broken by a high-velocity impact or torn up by gravitational effects.

Why is Uranus tipped on its axis? _____

_____

Answer: Possibly early in its history Uranus suffered a collision that knocked it over.

9.19 URANUS'S MOONS

Five large moons and at least 27 small moons orbit Uranus (Table 8.3).

The biggest moons look like tiny bright dots through our large telescopes. Titania was the first discovered, in 1787, and Miranda the last, in 1948. Voyager 2 found that the moons are dark gray ice-rock conglomerates, appar-

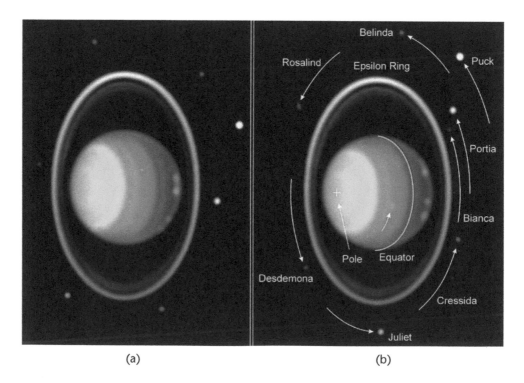

(a)   (b)

Figure 9.25. Uranus and some small moons from Hubble Space Telescope. Moons of Uranus are named for characters in writings of William Shakespeare and Alexander Pope.

ently made of about 50 percent water ice, 20 percent carbon and nitrogen-based materials, and 30 percent rock.

Miranda, the smallest of the five, looks strangest. It has huge fault canyons as deep as 20 km (12 miles), terraced layers, a chevron feature, large relief mountains, ridges, and rolling plains. This mixture of different terrain types on older and younger surfaces suggests diversity of tectonic activity, violent impacts, and tidal heating caused by Uranus's gravitational tug.

The two largest moons, Titania and Oberon, are about half the size of our Moon. Ariel has the brightest and possibly youngest surface, with many fault valleys and apparent extensive flows of icy material. Titania has huge fault systems and canyons that evidence past geologic activity. The surfaces of darkest Umbriel and Oberon look heavily cratered and old, indicating little past geologic activity.

Ten small moons were discovered by Voyager 2. Puck, the biggest, is 155 km (96 miles) in diameter. They are made of more than half rock and ice. Portia and Rosalind are shepherding the outermost ring, Epsilon, apparently keeping it in a narrow region. The Hubble Space Telescope and superior ground-based telescopes spotted the other moons (Figure 9.25).

(a) How do the surfaces of Ariel and Umbriel differ? _____

_____

(b) What do these differences indicate? _____

_____

Answer: (a) Ariel is bright with many fissures and apparent extensive flows of icy material; Umbriel is dark, heavily cratered, and old. (b) Ariel: geologic activity; Umbriel: little geologic activity.

9.20 PLANET NEPTUNE

When Voyager 2 flew within 5000 km (3000 miles) of **Neptune** in 1989, the planet was the most distant one from the Sun. The 8000 images Voyager sent back gave us our first good look at the Neptunian system (Figure 9.26). Like Uranus, Neptune has a thick hydrogen, helium, and methane cloud cover that appears bright blue.

Neptune's discovery was a triumph for theoretical astronomy. Uranus did not follow the path Newton's law of gravity predicted it should. Astronomers John Adams (1819–1892) in England and Urbain Leverrier (1811–1879) in France calculated that its motion was being disturbed by another planet's gravity. They predicted where that unknown planet should be in the sky.

In 1846 astronomer Johann Galle (1822–1910) at the Berlin Observatory in Germany pointed to the predicted spot and found Neptune. The planet was named for the Roman god of the sea.

Although Neptune, the smallest planet of the gas giants, receives only 3 percent as much sunlight as Jupiter, it has a dynamic atmosphere. There the strongest winds on any planet blow westward, opposite the direction of rotation. Several large, dark spots and high, long, bright clouds, streaks, and plumes appear.

Great Dark Spot 1989 was a giant storm the size of Earth, which resembled Jupiter's Great Red Spot. It circuited Neptune every 18.3 hours. Winds blew up to 2000 km (1200 miles) per hour nearby. Great Dark Spot 1989 vanished, and a new northern one, Great Dark Spot 1994, was photographed by the Hubble Space Telescope.

Neptune's magnetic field is highly tilted, 47° from the rotation axis. It may be characteristic of flows in the interior. The magnetic field causes radio emission and weak auroras.

Voyager found four rings circling Neptune. They are so diffuse and the material in them is so fine that they could not be fully resolved from Earth.

Figure 9.26. Neptune from Voyager 2, computer enhanced to bring out small features. Bright clouds near Great Dark Spot 1989 changed their appearances in a few hours.

Why was Neptune's discovery a triumph for theoretical astronomy? _____

_____

_____

Answer: Theory predicted that an unseen planet must exist. Neptune was discovered by looking for it at the spot in the sky predicted by theory.

9.21 NEPTUNE'S MOONS

Neptune has at least 13 confirmed moons (Table 8.3).

Triton is the largest and the most interesting (Figure 9.27). Voyager data showed that Triton's surface has methane ice. Recent infrared measurements showed carbon monoxide and carbon dioxide and carbon dioxide ices as well. Active geyserlike eruptions shoot invisible nitrogen gas and dark dust particles up several kilometers.

Figure 9.27. Neptune-facing hemisphere of the planet's largest moon, Triton, in a composite of about 12 images from Voyager 2.

The surface temperature is the coldest observed in the solar system, about 38 K (–391° F). The large south polar cap is slightly pink. From the ragged edge northward Triton is darker and redder, possibly colored by ultraviolet light and magnetospheric radiation acting on its atmospheric and surface methane.

A very thin atmosphere extends some 800 km (500 miles) above Triton's surface. Surface pressure is about 14 microbars, or 1/70,000 of Earth's. Nitrogen ice particles may form thin clouds a few kilometers above the surface.

Six small, dark moons discovered by Voyager 2 and five by enhanced ground-based telescopes remain close to Neptune's equatorial plane. They are named for mythology's water gods and nymphs. Proteus, the biggest, is 420 km (250 miles) in diameter. Small moons and rings are probably fragments of larger moons shattered in collisions.

With a relatively high density and retrograde orbit, Triton does not seem to be an original member of Neptune's family. Suggest a possible origin.

Answer: Perhaps Neptune captured Triton from an originally eccentric orbit.

9.22 DWARF PLANETS

Pluto, **Eris**, and **Ceres** were the first dwarf planets in this new category of solar system objects defined in 2006 by the International Astronomical Union (Section 8.1).

Pluto, named for the Greek god who ruled over the dead underground, was formerly a ninth planet. After astronomers saw bigger Eris and other similar Kuiper Belt objects, they reclassified it, calling all like transneptunian (beyond Neptune) dwarf planets **plutoids**. Asteroid Ceres (Section 8.15) was reclassified because it has about one-third the mass of the asteroid belt, seems almost spherical, and has not cleared its neighborhood.

Eris, in 2003, and its moon Dysnomia, in 2005, were discovered by U.S. astronomers Michael Brown and his team. They were named amidst intense controversy over how to classify newly discovered or understood objects: Eris, for the Greek goddess of discord and strife, and Dysnomia, for Eris's daughter, spirit of lawlessness.

Both Pluto and Eris have very eccentric and inclined orbits. Pluto moved inside Neptune's orbit in 1980, reached perihelion at 30 AU in 1989, and crossed outbound toward aphelion at 50 AU in 1999. Eris revolves from 38 AU to 97 AU.

Pluto, about 2300 km (1430 miles) in diameter, is two-thirds as big and one-sixth as massive as Earth's Moon. Eris is about 2400 km (1490 miles) in diameter. Low densities indicate that both must be made of ice and rock.

A thick layer of water ice and frozen methane, nitrogen, and carbon monoxide, polar caps, and large, dark spots near the equator cover Pluto's surface.

Apparently a layer of frozen methane covers Eris's. Mantles of water ice may surround rocky cores. Exterior ices thaw and form thin atmospheres when closer to the Sun and refreeze farther out in the coldest parts of the orbits.

Pluto has at least five moons. They likely formed when the dwarf planet collided with another object early in the solar system's history. The smashup hurled debris from both, which coalesced into moons. Charon is largest, about half the size of Pluto. It was named for the mythological boatman who ferried souls of the dead to Hades. Charon orbits Pluto in 6.387 days, exactly the time both take to rotate once. So an astronaut on Pluto would always see Charon in the same spot in the sky and only from one hemisphere.

Most dwarf planets are in the Kuiper belt. The IAU recognized Haumea in 2005 and Makemake in 2008. U.S. New Horizons (2006–), the first robot to Pluto and the Kuiper Belt, will add more. ▶http://pluto.jhuapl.edu◀

Refer to Figure 9.28. Can you explain how astronomers first verified that Pluto is not a star? _____

_____

_____

_____

Answer: Photographs of Pluto taken at different times show how it "wanders" relative to the background stars. (U.S. astronomer Clyde Tombaugh discovered Pluto by painstakingly comparing millions of star images and planet suspects on pairs of photographs of sections of the sky taken on different dates.)

Figure 9.28. Two photographs of Pluto, showing motion of the planet in 24 hours. Pluto is very faint, with its brightest magnitude +14.

SELF-TEST

This self-test is designed to show you whether or not you have mastered the material in Chapter 9. Answer each question to the best of your ability. Correct answers and review instructions are given at the end of the test.

1. Match each planet to a famous feature visible in a small telescope.

   ____ (a) Phases.   (1) Mars.
   ____ (b) Polar ice caps.   (2) Jupiter.
   ____ (c) Great Red Spot.   (3) Saturn.
   ____ (d) Rings.   (4) Venus.

2. Match common features to correct planet pairs.

   ____ (a) Alternate, parallel, dark   (1) Mercury and Venus.
             and light cloud bands.   (2) Jupiter and Saturn.
   ____ (b) Many craters and mountains.   (3) Uranus and Neptune.
   ____ (c) Thick hydrogen, helium, and
             methane cloud covers.

3. List three reasons why Venus would be a very unpleasant planet to visit.

   (1) _____

   (2) _____

   (3) _____

4. Figure 9.29 shows Venus, Earth, and Mars in orbit around the Sun. What letter in the diagram indicates the following points?

   (1) Venus is an evening star. _____
   (2) Venus is in a new phase. _____
   (3) Mars is at opposition. _____
   (4) Mars is not visible in our nighttime sky. _____

5. Sketch and label the three principal layers of the Earth.

   (1) _____ ; (2) _____ ; (3) _____

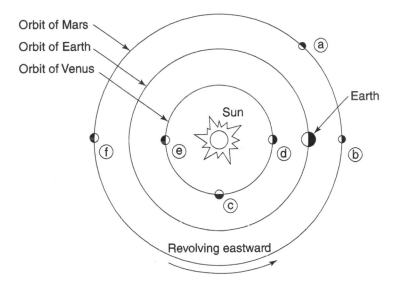

Figure 9.29. Aspects of Venus and Mars from Earth.

6. Give three observations that support the theory of plate tectonics (continental drift).

   (1)_____

   _____

   (2)_____

   _____

   _____

   (3)_____

   _____

7. Describe the scene, atmosphere, and temperatures at the robot landing sites on Mars. _____

   _____

   _____

8. State two observations that indicate water may have flowed on Mars long ago.

(1) _____

_____

(2) _____

_____

9. List the most abundant gases in the atmospheres of

(a) Earth _____

(b) Mars _____

(c) Jupiter _____

(d) Saturn _____

(e) Uranus _____

(f) Titan _____

10. List two observations that support the decision to classify Pluto as a dwarf planet.

(1) _____

(2) _____

11. Match a planet's moon to:

_____ (a)   Largest moon in solar system.

_____ (b)   Only moon known to have a substantial atmosphere.

_____ (c)   Most geologically active moon, with active volcanoes.

_____ (d)   Strangest, with mixture of young and old surfaces.

_____ (e)   Coldest surface, with active geyserlike eruptions.

(1) Ganymede/Jupiter.

(2) Io/Jupiter.

(3) Miranda/Uranus.

(4) Titan/Saturn.

(5) Triton/Neptune.

ANSWERS

Compare your answers to the questions on the self-test with the answers given below. If all of your answers are correct, you are ready to go on to the next chapter. If you missed any questions, review the sections indicated in parentheses following the answer. If you missed several questions, you should probably reread the entire chapter carefully.

1. (a) 4; (b) 1; (c) 2; (d) 3. (Sections 9.2, 9.9, 9.13, 9.14, 9.16, 9.18, 9.20)

2. (a) 2; (b) 1; (c) 3. (Sections 9.1, 9.3, 9.13, 9.14, 9.16)

3. Poisonous carbon dioxide atmosphere; much too hot (up to 900° F); and a crushing atmospheric pressure over 90 atmospheres. (Section 9.3)

4. (1) c; (2) d; (3) b; (4) f. (Sections 9.2, 9.9)

5. Figure 9.12: (1) crust; (2) mantle; (3) core. (Section 9.5)

6. (1) Similar plant and animal fossils are found along coastlines of South America and West Africa. (2) These coastlines seem to fit together. (3) No rocks from the bottom of the Atlantic Ocean near the coastlines are older than about 150 millions years. (Section 9.6)

7. The surface looks like a red, dry, rock-strewn desert. The sky is pink. The temperature is cold. (Section 9.10)

8. (1) Deep, winding channels that look as if they were carved by great rivers; (2) water frozen in the polar ice caps. (Section 9.11)

9. (a) Nitrogen (about 78 percent) and oxygen (about 21 percent); (b) carbon dioxide; (c) hydrogen and helium; (d) hydrogen and helium; (e) hydrogen and helium, with some methane; (f) nitrogen. (Sections 9.8, 9.11, 9.14, 9.16 through 9.18)

10. (1) Pluto's properties do not fit into either the terrestrial or giant planet category. (2) Many similar icy worlds are orbiting with Pluto in the Kuiper Belt. Eris is larger, and astronomers expect to find more large KBOs. (Section 9.22, Table 8.3)

11. (a) 1; (b) 4; (c) 2; (d) 3; (e) 5. (Sections 9.15, 9.17, 9.19, 9.21)

10

THE MOON

The stars about the lovely Moon hide their shining forms when it lights up the Earth at its fullest.

Sappho (c. 612 B.C.)

Fragment 4

Objectives

☆ Explain the Moon's appearance and apparent motions in the sky.

☆ Compare the Moon and Earth in diameter, mass, average density, and surface gravity.

☆ Describe the general surface features of the Moon.

☆ Compare and contrast the Moon and Earth with respect to geological activity and erosion of surface features.

☆ Outline a hypothesis of the origin of the Moon that is consistent with observations.

☆ Explain the probable origin of lunar craters and maria.

☆ Describe surface conditions on the Moon at the Apollo landing sites.

☆ Give the current model of the Moon's internal structure.

☆ List some questions about the Moon that remain to be answered.

☆ Describe the relative positions of Earth, Moon, and Sun during a solar eclipse and a lunar eclipse.

10.1 NEAREST NEIGHBOR

Poets have always been enchanted by the beautiful Moon. At magnitude –12.5, a full Moon is almost 25,000 times more brilliant than first-magnitude stars (Figure 10.1).

Figure 10.1. The full Moon. The six sites where U.S. Apollo astronauts landed are marked 11 through 17 (corresponding to mission numbers). Prominent maria, craters, and mountain ranges are named on your Moon Map.

People once believed that the brilliant Moon influenced personal behavior directly. They practiced special rituals at full Moon. Some of the ancient names for a Moon goddess were Diana, Lunae, Selene, and Cynthia. Words such as "Moonstruck" and "lunacy" originally referred to a madness that changed with the phases of the Moon.

Today we know more about the Moon than about any other neighbor in space. It is the closest sky object of all, located an average of 384,400 km (239,000 miles) from Earth. Robot spacecraft and astronauts have been to the Moon and returned thousands of photographs, scientific data, and surface samples.

U.S. **Apollo Moon missions** (1969–1972) landed men with cameras and equipment for scientific experiments and returned 382 kg (842 pounds) of Moon rock for laboratory study. ▶http://nssdc.gsfc.nasa.gov/planetary/lunar/apollo.html◀ Apollo instruments sent back data until 1977. They were finally turned off for budgetary reasons.

The Moon shines by reflecting sunlight. Its average visual **albedo**, the proportion of incident sunlight that the Moon reflects back into space, is only 11 percent. Most of the sunlight that shines onto the airless Moon's surface is absorbed.

Why do you think the full Moon is the brightest light in the night sky?

Answer: Because it is so much closer to Earth than all other sky objects.

10.2 IN THE SKY

If you look at the Moon regularly, you will observe its two apparent motions in the sky in addition to its phases (Section 8.4).

You will see the Moon rise in the east, move westward across the sky, and set every day, because the Earth rotates daily.

You will also observe that the Moon changes its location with respect to the stars about 13° to the east every day, because the Moon moves with respect to the Sun daily as the Earth-Moon system revolves around the Sun every year (Figure 10.2).

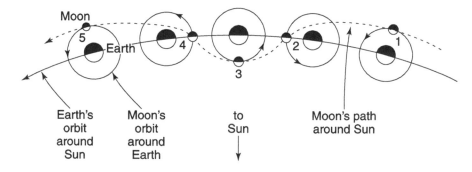

Figure 10.2. The Earth-Moon system's revolution around the Sun. The waviness of the Moon's orbit is greatly exaggerated for clarity.

Explain why the Moon rises an average of about 50 minutes later each day than it did the day before. _____

_____

_____

_____

Answer: At Moonrise the Moon is located in a particular constellation. About 24 hours later, when Earth has turned completely around, those same stars rise again, but meanwhile the Moon has moved about 13° eastward with respect to the stars and so does not rise until later.

10.3 ROTATION

Earth's gravity has locked the Moon into **synchronous rotation**. The Moon rotates on its axis every 27.3 days, the same amount of time it takes to travel around Earth. That makes the same side of the Moon face Earth at all times (Figure 10.3).

Notice that you observe the same features of the "man in the Moon" all month long, but never see the back of his head. (The visible disk of the Moon appears to shift, due to slight variations in the Moon's motions, called **libration**. You can actually see a total of 59 percent of its surface over time.)

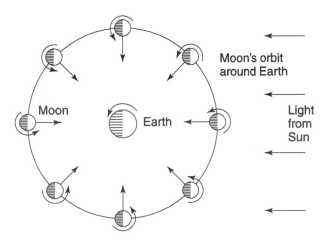

Figure 10.3. The Moon's synchronous rotation. Note: The same side of the Moon always faces Earth.

The Moon's rotation period and its revolution period are probably not equal by coincidence, but are equal because of eons of tidal friction.

Why were people able to see only one side of the Moon before spacecraft flew to the far side? _____

_____

Answer: The Moon's rotation period is equal to its period of revolution around Earth, so the same side of the Moon always faces Earth.

10.4 SPECIAL EFFECTS

You may observe other dramatic changes in the Moon's appearance.

A **lunar halo**, or ring around the Moon, is really not near the Moon at all. Ice crystals high up in Earth's atmosphere refract moonlight as it passes through, creating the halo effect.

When the Moon is near the horizon, it may look red. From that position, moonlight travels a longer path through the atmosphere to our eyes than when the Moon is overhead. Moonlight (reflected sunlight) consists of all visible colors. Short (blue) moonlight rays are scattered out and the long (red) rays, which penetrate the atmosphere more readily, color the Moon red.

A full Moon looks bigger when it is near the horizon than when it is high in the sky. The size of the Moon is always the same. No one knows exactly what causes this common **Moon illusion**. Perhaps comparing the Moon with local objects as opposed to distant stars makes the difference psychologically.

The **harvest Moon** is the full Moon nearest the time of the autumnal equinox. It rises earlier in the evening than usual, lighting up the sky to give farmers extra hours for harvesting. Harvest Moon occurs when the angle between the ecliptic and the horizon is near minimum.

Do you think the Apollo astronauts saw a "ring around the Earth" from the Moon's surface? Explain. _____

_____

_____

Answer: No. The ring around the Moon is an illusion caused by ice particles in Earth's atmosphere. The Moon has no atmosphere or water to create the illusion of a ring around Earth in the sky.

10.5 SIZE

The Moon is unusually large for a satellite, in comparison to its parent planet. The Moon's size can be found from measurements of its angular diameter and its distance from Earth.

The distance to the Moon has been measured to a fantastic accuracy of one part in 10 billion (a few centimeters) by timing how long it takes a laser light beam to reach reflectors there and return.

The equatorial diameter of the Moon is 3476 km (2160 miles). The equatorial diameter of Earth is 12,756 km (almost 8000 miles).

Compare the size of the Moon to that of Earth. _____

Answer: The diameter of the Moon is about ¼ that of Earth.

Solution: Diameter of Moon/diameter of Earth ≅ 3500 km/13,000 km ≅
(2000 miles/8000 miles) = ¼.

10.6 DENSITY

The Moon's mass, measured from the accelerations the Moon produces on spacecraft, is 7.35 x 10²² kg, or ⅛₁ that of Earth.

The Moon's average density is 3.34 t/m³, or roughly ⅗ that of Earth.

The Moon's surface gravity is only about ⅙ that of Earth because of its small mass and size. That means an 84-kg (180-pound) astronaut would weigh only 14 kg (30 pounds) on the Moon's surface.

Suggest a reason why the Moon's average density is less than Earth's.

_____

_____

_____

Answer: The Moon must be made almost entirely of silicate rocks like those of Earth's crust and mantle, and be poor in iron and other metals. (Moon rocks analyzed so far are made of the same chemical elements as Earth rocks, but the proportions are different.)

10.7 DATA

Review the properties of the Moon you have learned so far by completing Table 10.1. Review Sections 8.4 and 8.10, if necessary.

TABLE 10.1 Properties of the Moon

Quantity	Value
(a) Average distance from Earth	_____
(b) Diameter	_____
(c) Sidereal orbital period (fixed stars)	_____
(d) Synodic orbital period (phases)	_____
(e) Rotation period	_____
(f) Mass	_____
(g) Average density	_____
(h) Surface gravity	_____
(i) Albedo	_____
(j) Apparent magnitude of full Moon	_____
(k) Average velocity in orbit	_____

Answer: (a): 384,400 km (239,000 miles); (b) 3476 km (2160 miles), or ¼ that of Earth; (c) 27.3 days (27.321 66 days); (d) 29.5 days (29.530 59 days); (e) 27.3 days; (f) 7.35 x 10^{22} kg, or 1/81 that of Earth; (g) 3.34 t/m³, or ⅗ that of Earth; (h) ⅙ that of Earth; (i) 0.11; (j) –12.5; (k) 1.02 km/sec (2295 miles per hour).

10.8 OBSERVING 🌘

The Moon has long been a favorite target for binoculars and low-power telescopes because it is close enough to see in great detail.

The Moon Map at the back of this book has been drawn especially to make it easy for you to identify prominent surface features. It shows the Moon as it appears around its highest in the sky in the northern hemisphere. Compass directions on your Moon Map correspond to sky directions. This map is oriented with north at the top, the way we see the Moon with our eyes or through binoculars.

(Through many telescopes the Moon appears inverted, with north at the bottom. For astronauts on the Moon or topographic maps of the surface, east and west are interchanged as with Earth maps; north and south remain unchanged.)

When Galileo first pointed his telescope at the Moon, he mistakenly thought the large, relatively smooth dark areas he saw were oceans. He called them **maria** (singular **mare**), meaning "seas."

Maria are actually dry lava beds made of **basalt**, a dark, dense igneous rock. They formed over three billion years ago when molten lava from the

Moon's hot interior flooded huge impact basins. Mare Imbrium, the Sea of Showers, the largest mare of this type, is about 1100 km (700 miles) across.

The brighter areas of the Moon are called **highlands**. They are higher, more rugged, older regions than the maria. Highlands contain light-colored igneous rocks. They cover about 80 percent of the surface.

What are the maria that form the features of the "man in the Moon"?

Answer: Solidified lava beds.

10.9 CRATERS

The Moon is pitted with **craters**, or holes in the surface.

The craters are customarily named after famous scientists and philosophers such as Copernicus and Plato. Large, flat-floored craters—such as Clavius, nearly 240 km (150 miles) across—are called walled plains. The smallest are known as craterlets.

Typical craters are circular, ranging in size from tiny pits to huge basins. High-speed meteorites probably blasted out most on impact (Figure 10.4). The largest, far side South Pole-Aitken basin, is nearly 2600 km (1500 miles) across and 8 km (5 miles) deep.

The heat of impact vaporizes a meteorite and some of the ground it penetrates. The hot, vaporized material expands violently and explodes, forming a circular crater with a high rim and often a central peak. Ejected material falls back around the crater, often forming smaller secondary craters.

Bright **a ys** radiate for hundreds of kilometers across the surface from

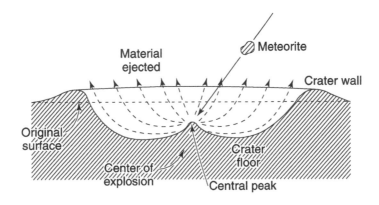

Figure 10.4. Formation of a typical impact crater.

young prominent craters. These are apparently splash patterns made of ejected material from impact explosions.

The best time to observe craters and mountains is when they are near the sunrise or sunset line, called the **terminator**. Then the Sun's low elevation above the ground causes shadows that highlight surface relief.

The sunrise terminator moves from right to left over the Moon's surface between its new and full phases. The sunset terminator does the same between full and new Moon. When the Moon is full, the maria stand out prominently, but the lack of surface shadows makes the surface relief hard to discern.

Spacecraft photographs show that the far side of the Moon has craters and highlands, but it does not have large maria, which are so conspicuous on the near side. Apparently the far side's thicker outer layer prevented upwelling lava from pouring into the basins.

What probably produced most craters on the Moon?_____

Answer: Meteorites crashing into the surface.

Figure 10.5. The far side of the Moon was first photographed by Russian robot Luna 3 on October 4, 1959.

10.10 EXPLORING

When U.S. astronaut Neil Armstrong (1930–2012) first set foot on the Moon for "all mankind" on July 20, 1969, he entered a strange, desolate world.

The entire surface of the Moon is covered by powdery soil called **regolith**, which is produced by the shattering of surface rock during prolonged meteorite bombardment. Many areas have a lot of loose rock ranging in size from pebbles to huge boulders.

No water flows, nothing grows. No fossils or organisms of any kind have been found in laboratory analysis of Moon rocks and soil. This lack of evidence of life suggests that the Moon is and has always been lifeless.

No blue sky, white clouds, or weather of any kind appear above the Moon because the Moon's atmosphere is extremely thin. Ghostly silence prevails in the absence of air to carry sounds.

Figure 10.6. Twelve U.S. Apollo astronauts spent a total of 300 hours on the Moon. The last three pairs drove lunar rovers and explored the surface around their landing sites.

Days and nights are long—14 Earth-days each. The surface temperature at the equator ranges from about 120° C (250° F) when the Sun is at its highest point to about –150° C (–240° F) at night.

Apollo astronauts proved that the Moon is accessible for human activity (Figure 10.6). In the twenty-first century, astronauts at a Moon base could perform astronomical and other scientific research and also process resources such as oxygen and metals to support new space exploration.

Why would it be useful to place a large optical telescope on the Moon?

_____

_____

Answer: In the absence of air or weather on the Moon, the seeing (Section 2.21) would always be good.

10.11 BOMBARDMENT

The major agents of erosion of the Moon's surface are **micrometeorites**, tiny grains of rock and metal, that crash into the Moon at speeds of up to 113,000 km (70,000 miles) an hour. Large meteorites also collide with the Moon occasionally.

Micrometeorites are about 10,000 times less effective in changing the Moon's surface than are air and water on Earth. They remove only about 1 millimeter of lunar surface in a million years.

Explain why Neil Armstrong's first boot print on the Moon will probably look the same millions of years from now as it did in 1969 (Figure 10.7).

_____

_____

Answer: Erosion on the Moon is due primarily to micrometeorite bombardment and happens much more slowly than erosion by air and water on Earth.

10.12 MOUNTAINS

Mountains on the Moon are named after the great ranges on Earth, such as the Alps. They are different from ours in both chemical composition and appearance because they were produced and shaped by different forces.

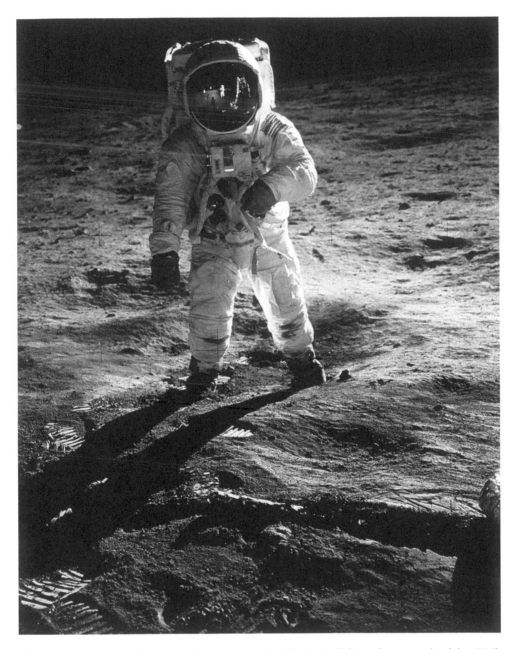

Figure 10.7. Historic Apollo 11 Moon visit. Edwin E. Aldrin photographed by Neil Armstrong, whose reflection was captured in the helmet visor.

The highest rugged mountain peaks on the Moon tower over 9000 m (29,000 feet), as does Mount Everest, Earth's highest mountain.

What two major factors that constantly change the shape of Earth's mountains do not shape Moon mountains? Explain. _____

_____

Answer: Water and atmosphere. No mountain streams pour down these ranges. No atmospheric storms ever rage to wear away the surface.

10.13 HISTORY

Astronomers reconstruct this life story of the Moon from direct lunar exploration.

The oldest Moon rocks, collected in the highlands, are about 4.3 billion years old. A few tiny green rock fragments are about 4.6 billion years old. The youngest, from the maria, were formed about 3.1 billion years ago.

The manner and place of birth of the Moon remain a mystery.

Moon rocks are richer in silicates and poorer in metals and volatile elements than Earth rocks. So the Moon was probably not originally a part of Earth that was torn off. It probably did not form by the accretion of many smaller particles in the solar nebula, either.

The popular Giant Impact theory says that soon after Earth formed, a planet-sized body crashed into it off-center. The impact ejected a huge, hot glob of material from Earth and impactor, which broadened to a ring around our planet. Material in the ring then collected to form the Moon.

During its first billion years of life, the young Moon was heavily bombarded by meteorites of all sizes. They produced great craters and melted the surface, now the lunar crust.

When the Moon was about 1 billion years old, the interior was greatly heated by radioactive elements. Volcanoes poured huge floods of hot basaltic lava over the surface and into the craters. This molten lava solidified, forming the maria.

About 3 billion years ago the Moon cooled off significantly, and volcanic activity largely stopped. Except for minor lava flows and a relatively small number of large impact craters like the young (about 1 billion years old) Copernicus, the Moon has not changed since. Seismographs left on the Moon by Apollo astronauts detected a very low level of Moonquakes.

The Moon's airless, dry, stable surface preserves a historical record of ancient bombardments that must have been common to all terrestrial planets.

How does the story of the Moon's geological activity differ from Earth's?

_____

_____

_____

Answer. The Moon became essentially dead geologically after the first 2 billion years of its life in contrast to Earth, which is still very much alive with volcanism, mountain building, and drifting continents.

10.14 INSIDE THE MOON

Geologists draw their current picture of the Moon's interior from spaceflight data. Gravity field measurements revealed **mascons**, mass concentrations, submerged in the circular maria. The existence of mascons plus the absence of major Moonquakes suggest that the Moon has a cold, thick, rigid outer layer, or crust. The crust is about 60 km (40 miles) thick on the near side and thicker on the far side.

Beneath the crust, extending down about 1000 km (600 miles), is the mantle. The physical characteristics of the core, extending the last 700 km (400 miles) to the center, are still unknown. The core may be partly molten, at a temperature up to 1500 K.

The Moon does not have a sizable magnetic field today, but old lunar rocks indicate that it once did.

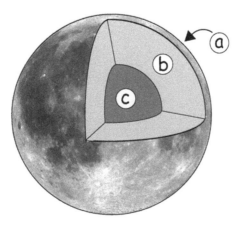

Figure 10.8. The Moon's structure.

Refer to Figure 10.8. Identify the crust, mantle, and core, and indicate the approximate depth of each layer. (a)_____

_____ ;

(b) _____ ;

(c) _____

Answer: (a) Crust: 60 km (40 miles) on the near side and thicker on the far side; (b) mantle: 1000 km (600 miles); (c) core: 700 km (400 miles).

10.15 SURFACE CONDITIONS

Questions about the Moon proliferate. New data show a scattering of ice in the regolith, frost deep inside polar craters, and a variable exosphere of unusual gases and structure. Research and exploration, as well as analysis of more lunar material is ongoing.

Summarize what we have learned about the surface of the Moon so far.

_____

_____

Answer: Your paragraph should describe the maria, craters, mountain ranges, absence of air and flowing water, length of day and night, and surface temperatures.

10.16 ECLIPSE OF THE SUN

A **solar eclipse** occurs when the Earth, new Moon, and Sun are directly in line (Figure 10.9).

The eclipse is **total** when the Moon is closer to Earth than the length of its shadow cone. The Moon looks bigger than the Sun and blocks the Sun's bright disk from view.

Totality lasts only a few minutes and can be seen only at successive places along a narrow curved path (a few hundred kilometers wide) inside the Moon's shadow on Earth. The maximum duration of totality is 7.5 minutes, which is not predicted to occur again until July 16, 2186.

Over a wider region bordering both sides of the path of totality a **partial eclipse** is seen. A partial eclipse may also occur when the Moon is not quite close enough to the Sun–Earth line to block all of the Sun from view (Figure 10.10).

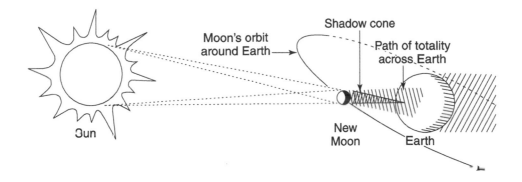

Figure 10.9. Solar eclipse (not to scale).

An **annular eclipse** occurs when the Moon is farther from Earth than the length of its shadow cone. The Moon looks smaller than the Sun and blocks from view all of the Sun's bright disk except for an outer ring (annulus) of sunlight (Figure 10.11).

It is thrilling to observe a total eclipse of the Sun! When the Moon passes in front of the bright Sun, an unnatural darkness spreads across the sky, the temperature drops, and stars and planets shine in the daytime sky. People once connected the Sun's disappearance with terrifying events. Today professional and amateur astronomers eagerly go around the world to observe a total solar eclipse and gather astronomical data.

Your chances of observing a total solar eclipse from your hometown are very small. Any one location on Earth averages only about 1 occurrence in 360 years. You might consider joining an eclipse expedition to view this spectacular natural event. Coming total solar eclipses are listed in Table 10.2. See NASA's solar eclipse site. ▶eclipse.gsfc.nasa.gov/solar.html◀

Figure 10.10. Partial solar eclipse.

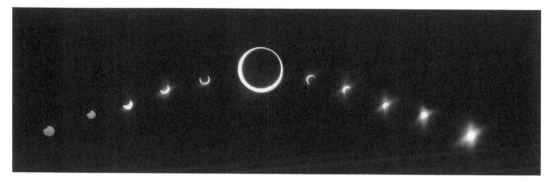

Figure 10.11. Annular solar eclipse.

TABLE 10.2 Total Solar Eclipses

Date	Duration of Totality (minutes:seconds)	Where Visible
2015 March 20	2:47	Arctic, North Atlantic
2016 March 9	4:09	Asia, Pacific
2017 August 21	2:40	North Pacific, U.S., South Atlantic
2019 July 2	4:33	South Pacific, Chile, Argentina
2020 December 14	2:10	Pacific, South America
2021 December 4	1:54	Antarctica
2023* April 20	1:16	Indonesia, Australia, Papua New Guinea
2024 April 8	4:28	Mexico, U.S., Canada
2026 August 12	2:18	Arctic, Spain

*__Hybrid eclipse__, or annular/total eclipse. The eclipse changes from total to annular or visa versa along different sections of the path when the curvature of Earth's surface causes a big enough variation in the Moon's distance.

What phase must the Moon be in for an eclipse of the Sun to occur?_____

Answer: New.

10.17 ECLIPSE OF THE MOON

A **lunar eclipse** occurs when the Sun, Earth, and full Moon are directly in line (Figure 10.12).

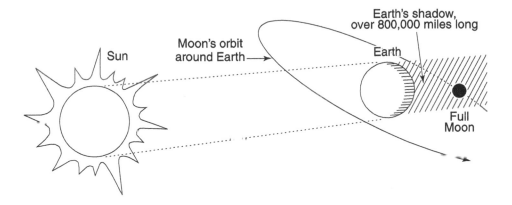

Figure 10.12. Lunar eclipse (not to scale).

The Moon darkens when it enters Earth's shadow, but it still gets some sunlight that is refracted around Earth by our atmosphere. Clouds, dust, and pollution affect the color and brightness of the Moon's appearance, usually making it dull red.

More than 2000 years ago, the Greeks observed that during a lunar eclipse, Earth's shadow appears circular on the Moon. Philosopher Aristotle (384–322 B.C.) cited this evidence to support the theory that the Earth is a sphere rather than flat. Astronomer Eratosthenes (c. 276–194 B.C.) made the first fairly accurate measurement of the diameter of the Earth.

Your chances of seeing a total lunar eclipse are much greater than your chances of seeing a total solar eclipse (Table 10.3). Lunar eclipses are safe to watch and can be seen from any place on Earth where the Moon is shin-

TABLE 10.3 Total Lunar Eclipses

Date	Duration of Totality (minutes)	Visible from North America
2015 September 28	72	Yes
2018 January 31	76	Yes
2018 July 27	103	No
2019 January 21	62	Yes
2021 May 26	75	Yes
2022 May 16	85	Yes
2022 November 8	85	Yes
2025 March 14	65	Yes
2025 September 7	82	No

ing. They last much longer than solar eclipses. The longest total eclipse of the twenty-first century is 1 hour, 43 minutes, predicted to occur on July 27, 2018. See NASA's lunar eclipse site. ▶http:eclipse.gsfc.nasa.gov/lunar.html◀

What phase must the Moon be in for a lunar eclipse to occur? _____

Answer: Full.

10.18 ECLIPSE TIMES

The maximum number of solar and lunar eclipses that can occur in one year is seven.

Eclipses do not occur every time we have a new or full Moon, as you might expect. The Moon's orbit is tilted 5.2° to the plane of Earth's orbit. Most months the Moon is above or below the Sun–Earth line at new and full phase, so no eclipse can occur (Figure 10.13).

The Moon's orbit crosses the plane of Earth's orbit at two points called **nodes**. The nodes move slowly westward, called **regression of the nodes**, because of the Sun's gravitational pull.

Examine Figure 10.14. Explain why an eclipse can occur only when the Moon is at point A or point B. _____

_____

Answer: Then the Sun, Earth, and Moon are directly in line.

Figure 10.13. Unfavorable conditions for an eclipse.

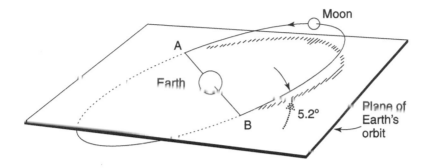

Figure 10.14. The plane of the Moon's orbit is tilted 5.2° to the plane of Earth's orbit.

10.19 OCCULTATIONS

Occultation means the eclipse of one sky object by another.

An occultation by the Moon is the most frequent and easiest type for you to observe. The Moon often passes between Earth and a star or planet, causing it to suddenly disappear and to reappear after the Moon moves by. Predictions of lunar occultations are published in current astronomical publications (see Useful References).

Jupiter is over 40 times bigger than the Moon. How is it possible for the Moon to occult, or hide, Jupiter?_____

_____

Answer: Jupiter is much farther away than the Moon, so it appears much smaller.

SELF-TEST

This self-test is designed to show you whether or not you have mastered the material in Chapter 10. Answer each question to the best of your ability. Correct answers and review instructions are given at the end of the test.

1. Why do earthbound observers always see the same side of the Moon?

   _____

   _____

2. The Moon is about what fraction of Earth in (a) diameter? _____

   (b) mass? _____ (c) average density? _____

   (d) surface gravity? _____

3. Match the lunar features with their names:

   ____ (a) Dry lava beds.

   ____ (b) Holes in the surface.

   ____ (c) Light-colored, higher, rugged older regions.

   ____ (d) Submerged clumps of mass.

   (1) Craters.
   (2) Highlands.
   (3) Maria.
   (4) Mascons.

4. Suppose you were leading an expedition to explore the surface of the Moon. Which of the following would be useful? (a) extra oxygen tanks; (b) flare pistol and flares; (c) flashlight; (d) magnetic compass; (e) matches; (f) star chart; (g) umbrella; (h) watch _____

   Explain. _____

   _____

   _____

5. What is the probable origin of most craters on the Moon? _____

   _____

   _____

6. Why do the Moon's surface features change so much more slowly than Earth's?_____

_____

_____

_____

7. How old are the (a) oldest and (b) youngest Moon rocks that have been collected? (a) _____ ;
(b) _____

8. Sketch the Moon's interior; identify its three main layers. (a) _____ ;
(b) _____ ; (c) _____

9. List three questions about the Moon that remain to be answered.

(1)_____

_____

(2)_____

_____

(3)_____

_____

10. What must the phase of the Moon be for (a) the occurrence of an eclipse of the Sun? _____ (b) an eclipse of the Moon? _____

ANSWERS

Compare your answers to the questions on the self-test with the answers given below. If all of your answers are correct, you are ready to go on to the next chapter. If you missed any questions, review the sections indicated in parentheses following the answer. If you missed several questions, you should probably reread the entire chapter carefully.

1. The Moon's rotation period and its period of revolution around Earth are equal, called synchronous rotation.   (Section 10.3)

2. (a) ¼ diameter;   (b) ¹⁄₈₁ mass;   (c) ⅗ density;   (d) ⅙ surface gravity.   (Sections 10.5 through 10.7)

3. (a) 3;   (b) 1;   (c) 2;   (d) 4.   (Sections 10.8, 10.9, 10.14)

4. (a); (c); (f); (h). Because the Moon has no significant air, liquid water, or sizable magnetic field, nothing that requires any of the others would be useful.   (Sections 10.8, 10.10, 10.14)

5. Meteorites crashing into the surface.   (Section 10.9)

6. There is no significant air or liquid water to cause erosion on the Moon as they do on Earth. There is no comparable geological activity, either. Micrometeorites crashing into the Moon are the major agents of erosion on the Moon's surface.   (Sections 10.11 through 10.13)

7. (a) About 4.6 billion years old;   (b) 3.1 billion years old.   (Section 10.13)

8. Refer to Figure 10.8. (a) Crust;   (b) mantle;   (c) core.   (Section 10.14, Figure 10.8)

9. What is the chemical composition of the surface at places away from the few Apollo and Luna landing sites? Are frozen water or other volatiles in the regolith and polar craters brought by meteorites or comets? How did the Moon originate? (You may have thought of others.)   (Sections 10.6, 10.13 through 10.15)

10. (a) New;   (b) full.   (Sections 10.16, 10.17)

11

COMETS, METEORS, AND METEORITES

When beggars die, there are no comets seen:
The heavens themselves blaze forth the death of princes.

William Shakespeare (1564–1616)

Julius Caesar, Act II, ii:30

Objectives

☆ State why comets and meteorites are of interest to scientists.

☆ Describe the current theory of the origin and composition of comets.

☆ Explain in terms of the current model of comet structure the changes in a comet's appearance as its distance from the Sun changes.

☆ Specify the relationship between comets and meteor showers.

☆ Distinguish between a meteoroid, meteor, and meteorite.

☆ Give the composition and probable origin of meteorites.

☆ List some possible effects on Earth of a major comet or meteorite impact.

11.1 COMETS

Bright comets have always fascinated people (Figure 11.1). Unlike ordinary stars, these fiery-looking objects appear and disappear unpredictably. Records of bright comets go back to the fourth century B.C. Throughout history people have dreaded them as omens of human disasters such as wars or famines.

Today we know that a **comet** is an icy member of our solar system. Comets travel in elliptical orbits around the Sun and follow the basic laws of physics. They are not supernatural signs at all.

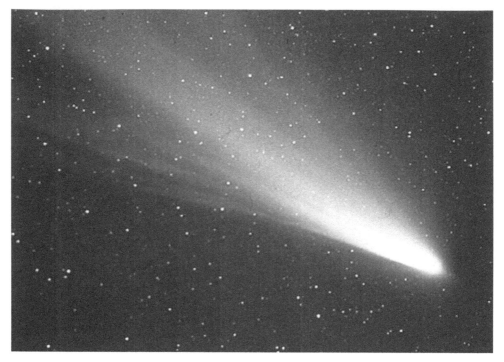

Figure 11.1. Famous Halley's Comet on March 16, 1986, five weeks after its most recent perihelion passage.

(a) What was the common historical view of comets?_____

_____

(b) What is the modern astronomical view of comets? _____

_____

Answer: (a) Comets were considered supernatural signs foretelling human misfortune. (b) Comets are icy members of the solar system that follow natural physical laws and have no hidden meaning.

11.2 SIGNIFICANCE OF COMETS

Comets that appear in our sky are important, even if they don't shine brilliantly. They are probably the only objects left that are made out of the original material from which the whole solar system formed about 4.6 billion years ago. Earth, Moon, and other celestial bodies have all been changed by tectonic processes, erosion, or numerous collisions. Only comets remain basically as they were in the beginning.

Robot spacecraft encounter and scrutinize comets with sophisticated tools (Table 11.1). Record firsts are European Space Agency (ESA) Giotto and U.S. Deep Space 1 / close-up images sent back in 1986 and 2001; U.S. Deep Impact / blasted out crater and studied ejecta in 2005; and U.S. Stardust / cosmic dust grains collected and returned to Earth in 2006. Next, ESA Rosetta is to orbit and drop a lander in 2014. ▶http://solarsystem.nasa.gov/missions◀

Why are comets important? _____

_____

_____

Answer: They are our best source for observing the original material out of which everything in the solar system was formed.

11.3 COMET STRUCTURE

Comets were named for their appearance. Both the Greek word *kometes* and the Latin word *cometa* mean "long-haired."

Figure 11.2. Main parts of a comet.

When it shines in the sky, a bright comet has a head with a starlike core called the **nucleus** surrounded by a glowing halo called the **coma** and long transparent **tails**. The nucleus is several kilometers in size. The coma may extend 100,000 km (60,000 miles) or more outside the nucleus. Tails can stream millions of kilometers into space.

Ultraviolet observations from spacecraft reveal a huge enveloping **hydrogen cloud**. It can grow up to tens of millions of kilometers wide. This cloud is not visible from Earth.

Refer to Figure 11.2. Identify the main parts of a typical bright comet.

(a) _____ ; (b) _____ ; (c) _____ ; (d) _____

Answer: (a) Nucleus; (b) coma; (c) tails; (d) hydrogen cloud.

11.4 THE NUCLEUS

Billions of comets probably orbit far out in the solar system, but you can't see them from Earth. They shine in the sky only when they travel near the Sun. The most widely accepted description of a typical comet is the **dirty snowball model**, proposed by U.S. astronomer Fred Whipple in 1950 (Figure 11.3).

When a comet is far out in the solar system it consists of only a nucleus. Its shape and surface are irregular. The nucleus is made of mostly water ice and other frozen gases (the "snow") loosely mixed with stony or metallic solids (the "dirt"). It has very low density and surface gravity.

Spacecraft images show that the icy nucleus is dark black and rotating. Surface irregularities include cracks, crevasses, and craters.

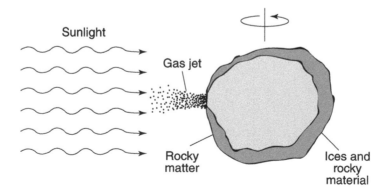

Figure 11.3. Dirty snowball comet model.

The nucleus becomes active as it advances through the inner solar system. Jets of dust and gas, primarily water vapor, erupt out of surface rifts whenever the nucleus faces the Sun. Many passages of a comet around the Sun drive off volatile gases, leaving most of the surface covered by a sooty black insulating dust layer.

Gases detected coming off the nucleus are 80 percent water vapor by volume and other compounds, including carbon dioxide, carbon monoxide, ammonia, and methane. Some dust grains seem to be silicates while others contain virtually only the elements carbon, hydrogen, oxygen, and nitrogen. These kinds of ices, dust, and gases form at low temperatures.

Some of the recently collected dust grains contain minerals that form at a high temperature. They must have formed near the Sun. How particles that originated in different environments ended up as aggregates in comets far from the Sun is puzzling.

Scientists are intrigued to find complex organic molecules in the collection, which could have significance for the origin of life on Earth.

What is the nucleus of a comet made of? _____

_____

_____

Answer: The dirty snowball model describes a comet nucleus as mostly water ice and other frozen gases loosely mixed with solids.

11.5 THE COMA

As a comet nucleus comes in from the edge of the solar system to within a few hundred million kilometers of the Sun, it heats up. Gases sublime and escape to space with dust from its surface. The comet's gravity is too weak to hold back the gases and dust. They expand outward around the nucleus for thousands of kilometers, forming the coma.

The comet shines because the gases fluoresce and the dust reflects sunlight. Astronomers use large telescopes to image about 25 of these blurry blobs of light each year.

What causes the coma to develop?_____

_____

Answer: The Sun's heat (which causes sublimation and expansion of gas and dust particles).

11.6 THE TAILS

When a comet moves near the Sun, it may develop tails of gases and dust released from the nucleus.

Ultraviolet radiation tears the gases apart into free radicals (molecular fragments) and ions. Ions interact with the charged particles blowing out from the Sun in the solar wind. The ions are ultimately swept millions of kilometers straight back into a **gas**, or **ion**, **tail**.

Radiation pressure, or intense sunlight striking, pushes the dust particles outward. The comet keeps moving, and a **dust tail** curves behind it. Comet tails are so thin you can see right through them to the stars on the other side.

Neutral molecules and atoms continue to expand outward until they are ionized. The most common atom, hydrogen, forms the huge hydrogen cloud. The hydrogen cloud surrounding the nucleus of Comet Halley at its 1986 apparition grew to several hundred thousand kilometers in diameter.

Effects of hydrogen ions released by Comet Halley on the solar wind were detected as far as 35 million km (21 million miles) from the nucleus. A **bow shock**, where cometary gases block and slow the solar wind, was found some 400,000 km (240,000 miles) in front of the comet.

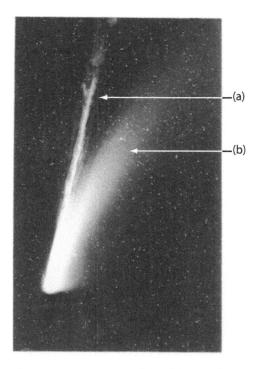

Figure 11.4. Comet Mrkos with typical tails.

Refer to Figure 11.4. Identify the gas or ion tail and the dust tail and state the cause of each. (a) _____ ;

(b) _____

Answer: (a) Ion tail; solar wind; (b) dust tail; radiation pressure.

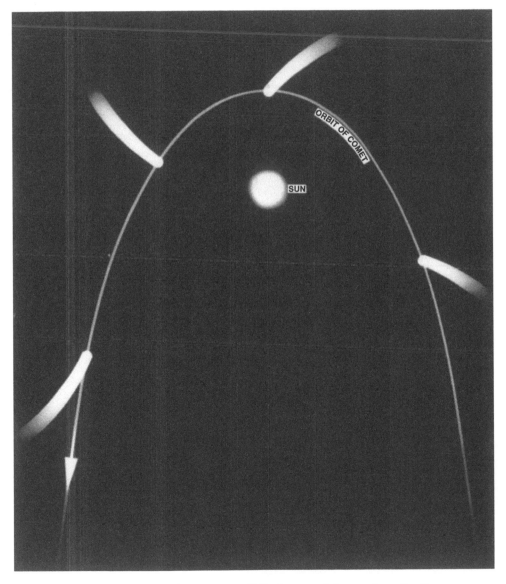

Figure 11.5. A comet's perihelion passage. Tails always point away from the Sun.

11.7 DISAPPEARANCE

As the comet accelerates inexorably nearer the Sun, its fate is unpredictable. Powerful jets of gas and dust from the nucleus may change its orbital motion.

If a comet rounds the Sun whole, it continues along its orbit back to frigid outer space. Some cometary material is left behind and the rest refreezes. Coma and tails vanish.

Some comets pass so close to the fiery Sun that they shatter or disintegrate. Occasionally one crashes directly into the Sun and disappears.

Refer to Figure 11.5. Why do comets go back to outer space tail first?

_____

_____

Answer: Comet tails are caused by solar radiation pressure and solar wind, which are always directed away from the Sun, so the tail must also point away from the Sun.

11.8 ORIGIN OF COMETS

"New" comets we observe apparently originate in a vast spherical shell of icy objects located 10,000 AU–100,000 AU from the Sun. This model was devel-

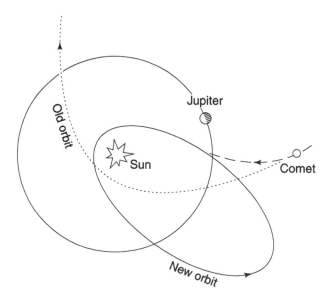

Figure 11.6. Jupiter's strong gravity perturbs a passing long-period comet into a new orbit around the Sun.

oped in the 1950s by Dutch astronomer Jan Oort (1900–1992). That unobserved **Oort cloud** may hold a trillion incipient comets.

Occasionally a passing star tugs on a comet, slows it down, and plunges
it randomly toward the Sun. That comet will be a **long-period comet**, with
an almost parabolic orbit and a period of revolution around the Sun of 200 to
millions of years.

Comets with shorter periods that orbit like the planets evidently originate
in the Kuiper Belt (Section 8.15).

If a comet passes near a giant planet, notably Jupiter, it will be affected
by that planet's strong gravity. Then the comet may crash into the planet, or
speed up and head out of the solar system, or move into an elliptical orbit
closer to the Sun (Figure 11.6).

Where do the long-period comets we discover probably originate?_____

Answer: In a vast cloud of comets near the edge of the solar system.

11.9 PERIODIC COMETS

Astronomers have catalogued about 300 periodic comets that have been
observed on more than one return to perihelion. They have well-known periods
of revolution around the Sun and shine in the sky every time they come close to
the Sun.

The most consistently bright and most famous is Comet Halley, with 30
consecutive perihelion passages recorded since 240 B.C. Sighted telescopically
for over three years before and after its February 9, 1986, perihelion passage,
Comet Halley is also the best-analyzed comet so far.

Table 11.1 lists some comets that have appeared several times in our sky.
What is the shortest known period of revolution of a comet? _____

Answer: 3.3 years (Comet Encke).

11.10 COMET FATE

A periodic comet cannot be reactivated to grow a new coma and tails indefinitely. Its nucleus loses a surface layer a few meters deep each time it rounds
the Sun. Dust and gas litter its orbit. Comet Halley leaves about 1 percent of
its mass behind during each perihelion passage (Figure 11.7).

Finally, a periodic comet will lose all its volatile material. Large chunks

TABLE 11.1 Some Periodic Comets

Comet	Period[a] (years)	Closest Approach to Sun (in AU)	Spacecraft Explorer
1P Halley	76.1	0.59	Giotto
2P Encke	3.3	0.34	—
9P Temple 1	5.5	1.5	Deep Impact
19P Borrelly	6.4	1.4	Deep Space 1
81P Wild 2	6.4	1.6	Stardust

[a]Period may change over time.

Note: Comet names can change, so the International Astronomical Union designates a periodic comet by P/, preceded by the period-comet number, assigned in the order in which the comet's periodicity was recognized.

Figure 11.7. Comet Halley on 7 different days as it receded from the Sun after its apparition in 1910.

and numerous small fragments of nonvolatile solids may survive. The debris continues to orbit around the Sun like tiny planets.

Summarize five changes of appearance that a comet undergoes as it travels in its orbit around the Sun. _____

_____

_____

Answer: 1. Far from the Sun, a comet consists of a nucleus of frozen gases and dust. 2. Coma forms as a comet approaches the Sun. 3. Close to the Sun, tails form. 4. After going around the Sun, much cometary material refreezes. 5. Far from the Sun again, coma and tails are gone.

11.11 COMET HUNT

Several new comets are discovered every year. Professionals find some among their data in observatories, and amateurs discover the others.

Comets are usually named for their discoverers. Exceptions, such as Halley, named for Edmond Halley (1656–1742), honor the people who first computed their orbits. The first three people to report seeing a new comet may have their names permanently attached to it. Since hunting comets is a popular international activity, we sometimes get real tongue twisters, such as the short-period (5.3 years) Comet 45P Honda-Mrkos-Pajdusakova!

How could a comet give you immortality? _____

Answer: Discover one, and it will carry your name.

11.12 INTERPLANETARY REMAINS

Countless bits and pieces of rock and dust, called **meteoroids**, populate the inner solar system.

Earth is surrounded by interplanetary dust. It is observed at infrared wavelengths. Meteoroids enter Earth's atmosphere continually. Astronomers collect them at high altitudes, from arctic ice sheets and from the ocean floor, for analysis in laboratories. These meteoroids are similar to dust grains ejected from the nucleus of Comet Halley (Figure 11.8).

Figure 11.8. Piece of dust from a comet, magnified 15,000 times.

What is a meteoroid? _____

Answer: A small solid object orbiting the Sun in space.

11.13 SHOOTING STARS

Did you ever make a wish on a "shooting star" or "falling star"? These light flashes are not stars at all. They are **meteors**, streaks of light created by meteoroids that are plunging through Earth's atmosphere at speeds up to 72 km (45 miles) per second. Air friction burns tiny particles when they are 60 to 110 km (40 to 70 miles) above Earth.

On any clear, dark night you can see an average of six meteors an hour flashing unpredictably in the sky. Meteors occur but are not visible during the daytime because the sky is too bright.

A large meteoroid creates an exceptionally brilliant meteor, called a fireball. On February 15, 2013, an asteroid 19 m (20 yards) wide entered Earth's atmosphere at high speed. A spectacular fireball was seen over a wide area in Russia. When it was near Chelyabinsk, the fireball exploded and broke up violently. The resulting powerful shock wave damaged buildings and sent glass flying, injuring over 1,500 people. About 6 kg (13 pounds) of large and small fragments and a 650 kg (1433 pound) rock that crashed through ice into a lake were recovered.

What is a "shooting star" or "falling star"? _____

_____   _____

Answer: A "shooting star" or "falling star," or meteor, is the streak of light that can be seen when a meteoroid is burned upon entry into Earth's atmosphere.

11.14 METEOR SHOWERS

On several predictable dates every year you may see meteors pour down from one part of the sky. This type of display is called a **meteor shower**. Meteor showers are associated with comets. They occur when Earth, moving along its orbit around the Sun, crosses a swarm of meteoroids left behind by an active comet (Figure 11.9).

In 1910, when Earth was about to travel right through the tail of Halley's Comet, people panicked. What would you expect to happen if Earth passed through a comet's tail? _____

Answer: A bright (but harmless) meteor shower.

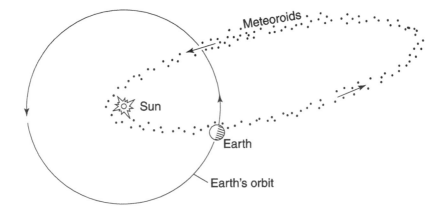

Figure 11.9. A meteor shower occurs when Earth passes near the orbit of a comet and encounters a swarm of meteoroids.

TABLE 11.2 Principal Annual Meteor Showers

Shower	Date of Maximum	Peak Activity Hourly Rate	Associated Comet
Quadrantids	January 2–3	30–80	2003 EH1
Lyrids	April 21–22	20	1861I G1 Thatcher
Eta Aquarids	May 5–6	10–45	1P Halley
South Delta Aquarids	July 28–29	20	Possibly 96P Macholz
Perseids	August 12–13	Up to 100	109P Swift-Tuttle
Orionids	October 21–22	20	1P Halley
Taurids	November 5	10	2P Encke
Leonids	November 17–18	15	55P Tempel-Tuttle
Geminids	December 13–14	120	Asteroid 3200 Phaêthon
Ursids	December 22	10	8P Tuttle

11.15 BEST METEOR DISPLAYS

During a meteor shower, all the meteors appear to come from one common point in the sky, called the **radiant** of the shower. Meteor showers are usually named for the constellation where they seem to originate, such as the Perseids from Perseus and the Orionids from Orion.

You can usually see more meteors after midnight than before, because Earth, moving along its orbit, is then traveling head-on into the swarms of particles. Meteor showers are best seen with your unaided eye on nights when the Moon is not bright. A full Moon can blot out a meteor shower.

Table 11.2 lists famous annual meteor showers. Shower activity varies, so it is best to consult ►http://www.jpl.nasa.gov/asteroidwatch/meteor.cfm◄ or a current astronomical publication (see Useful Resources) for details of this year's best showers.

Refer to Table 11.2. What is the name and date of maximum of the largest summer meteor shower observable from around 40° N latitude? _____

Answer: Perseids; August 12.

11.16 ROCKY LANDINGS

When a piece of stone or metal from outer space lands on Earth, it is called a **meteorite**.

Within recent history, no case has been documented of anyone being killed by a stone from the sky. Probably hundreds of tons of cosmic material

TABLE 11.3 Large Meteorites on Display in the U.S.

Meteorite	Approximate Weight	Present Location
Ahnighito (Greenland)	34 tons	American Museum of Natural History (New York City)
Willamette (Oregon)	14 tons	American Museum of Natural History
Furnas County (Nebraska)	1 ton	University of New Mexico
Paragould (Arkansas)	800 pounds	Chicago Natural History Museum

fall through the atmosphere every year, but only about two or three meteorites in a decade land in places where people live, and even more rarely is any related injury reported.

The largest meteorite ever found, the Hoba West, weighs an estimated 66 tons. It still lies in Namibia in Africa where it landed. A meteorite is usually named after the post office closest to its landing site. Many large meteorites are on display in museums worldwide (Table 11.3).

What is a meteorite? _____

Answer: A piece of stone or metal from outer space.

11.17 MAKEUP OF METEORITES

If you need extra spending money, find a meteorite! Scientists and collectors pay generously for genuine material from outer space.

Meteorites are divided into three main types: (1) **Iron meteorites** are about eight times as dense as water and consist mostly of iron (about 90 percent) and nickel. (2) **Stony-iron meteorites** are about six times as dense as water. They contain iron, nickel, and silicates. (3) **Stony meteorites** are about three times as dense as water. They have a high silicate content, and only about 10 percent of their mass is iron and nickel. Most are **chondrites**, distinguished by containing **chondrules**, or small silicate spheres.

Iron meteorites are found most often. Stony meteorites look like ordinary Earth rocks. They are not usually noticed unless they are seen falling. Laboratory analysis confirms their extraterrestrial origin. Table 11.4 lists the occurrence of the different types of iron and stony meteorites.

Most meteorites are probably fragments of asteroids shattered by colli-

TABLE 11.4 The Occurrence of Meteorite Types

Types	Meteorites Seen Falling	Meteorite Finds
Irons	6%	66%
Stony-irons	2%	8%
Stones	92%	26%

sions, since these objects have similar compositions. They appear to be about 4.6 billion years old, which is the approximate age of the whole solar system.

When a stony meteorite with a high carbon and some water content, called a **carbonaceous chondrite**, is found, it adds excitement to the search for life in space (Chapter 12). These finds show that prebiotic matter formed in the solar nebula when Earth did!

The 4.5-billion-year-old stony Murchison Meteorite fell in Victoria, Australia, in 1969. It contained simple amino acids, which make up proteins, and nucleic-acid bases that carry and replicate genetic information. It also had organic chemicals similar to lipids (structural components of living cells).

The Allende Meteorite fell in northern Mexico, also in 1969. It is one of the largest recorded carbonaceous chondrite falls. Almost 2 tons of some of the most primitive material in the solar system were recovered.

Collections of thousands of uncontaminated meteorites recovered from ice fields in Antarctica and the Sahara Desert in Africa include some carbonaceous chondrites containing amino acids.

Some "Moon" meteorites are close in composition to rocks collected on the Moon by the Apollo astronauts. A few "Mars" meteorites have gases trapped inside that are nearly identical chemically to the atmosphere of Mars. Probably a comet or asteroid crashed violently, ejecting chunks of rock that escaped their home to orbit through space, ultimately to be captured by Earth's gravity.

Why are meteorites important to scientists? _____

_____

_____

Answer: They are primitive extraterrestrial material that scientists can examine closely to determine more about the solar system.

11.18 IMPACTS ON EARTH

You may wonder what would happen if a comet or large meteorite struck Earth.

Large meteorites carve out huge craters in planets and moons when they crash. Earth must have been hit often by meteorites early in its history. Ancient craters are erased by geologic activity and erosion. Major impacts are now extremely rare. You can view the most recent, Meteor Crater near Winslow, Arizona, U.S. (Figure 11.10). It was created by a 100,000-ton iron-nickel meteorite impact over 50,000 years ago.

A comet nucleus could impact with the energy of millions of hydrogen bombs. Telescopes worldwide observed chunks of Comet Shoemaker-Levy 9 crash into Jupiter between July 16 and July 22, 1994. The impacts raised huge dark clouds high in the atmosphere that scarred Jupiter for months. Evidently the comet had been orbiting too close to Jupiter when tidal forces ripped it apart and shot its fragments into the planet.

Observations suggest that there are about 1000 near-Earth objects (Section 8.15) larger than 1 km (.6 mile) and a million larger than 100 m (330 ft) in diameter.

Figure 11.10. Meteor Crater in Arizona, U.S., is 1.2 km across and 180 m deep.

On June 30, 1908, in Siberia something exploded mysteriously about 8 km (5 miles) up in the air. The blast of about 12 megatons flattened over 1000 square km of forest near the Tunguska River. It killed reindeer 40 km (25 miles) away. A 100-m (330-ft) NEO could have blown up and caused the damage.

A major impact by a 15-km (9-mile) NEO with impact energy of 100 million megatons could have caused a catastrophic disruption of the biosphere and extinction of dinosaurs, many plants, and other animal species 65 million years ago. The huge Chicxulub crater buried under the Yucatan Peninsula in Mexico is the postulated impact site.

The site has enriched deposits of iridium in the K-T boundary, the geological layer of sediments laid down at the end of the Cretaceous Era and the start of the Tertiary Era 65 million years ago. Iridium is more abundant in comets, meteorites, and asteroids than in Earth's crust. Shock-melted mineral spherules and soot were also found in and near the K-T boundary.

Still most astronomers figure the chance of a direct big hit on Earth is remote. Most near-Earth objects are small. Interplanetary space is vast. None of the known big NEOs come dangerously near to Earth in their travels around the Sun. ▶http://neo.jpl.nasa.gov◀

Is it likely that Earth will be hit by a comet nucleus or large meteorite in the near future? _____

Answer: No.

SELF-TEST

This self-test is designed to show you whether or not you have mastered the material in Chapter 11. Answer each question to the best of your ability. Correct answers and review instructions are given at the end of the test.

1. Why do modern astronomers use their most sophisticated instruments to study comets? _____ _____

_____

_____

2. What is a comet nucleus made of? _____

_____

_____

3. State two important discoveries about the nucleus of comets that were made by robot spacecraft. (1)_____

_____

_____

(2)_____

_____

_____

4. Describe five changes of appearance that a periodic comet undergoes as it travels in its orbit around the Sun. _____

_____

_____

_____

_____

5. Sketch and identify the main parts of a typical bright comet.

    (a) _____ ; (b) _____ ;

    (c) _____ ; (d) _____

6. Describe the origin and extinction of long-period and periodic comets. _____

    _____

    _____

    _____

    _____

7. Match each description with the correct item.

    ____ (a) "Shooting star" or "falling star."   (1) Meteor.

    ____ (b) Small object orbiting the Sun.   (2) Meteorite.

    ____ (c) Solid body that reaches Earth.   (3) Meteoroid.

8. Explain the relation between comets and meteor showers.

    _____

    _____

    _____

9. Describe the composition and probable origin of most meteorites.

    _____

    _____

    _____

    _____

10. List the following in order of increasing distances from the Sun: asteroid belt, Earth, Oort comet cloud, Kuiper Belt._____

    _____

11. Explain why meteorites are of interest to scientists. _____

    _____

    _____

ANSWERS

Compare your answers to the questions on the self-test with the answers given below. If all of your answers are correct, you are ready to go on to the next chapter. If you missed any questions, review the sections indicated in parentheses following the answer. If you missed several questions, you should probably reread the entire chapter carefully.

1. Comets are considered to be the most unchanged objects left that are made out of the original material from which the whole solar system formed. (Section 11.2)

2. A comet nucleus is made of mostly water ice and other frozen gases loosely mixed with solids according to the dirty snowball model. (Sections 11.2, 11.4)

3. (1) The nucleus is dark black and rotating. (2) The surface has cracks, crevasses, and craters, a sooty black insulating dust layer, and jets of dust and gas escape near perihelion. (Section 11.4)

4. Far from the Sun, a comet consists of a nucleus of frozen gases and dust. Coma forms as a comet approaches the Sun. Tails form close to the Sun. After going around the Sun, a comet refreezes. Far away from the Sun, a comet consists of a nucleus again. (Sections 11.3 through 11.7, 11.9, 11.10)

5. Refer to Figure 11.2. (a) Nucleus; (b) coma; (c) tail; (d) hydrogen cloud. (Section 11.3)

6. Long-period comets probably originate in the vast Oort comet cloud near the edge of the solar system. Jupiter's strong gravity redirects those that pass nearby to elliptical orbits around the Sun. Periodic comets evidently originate in the Kuiper belt. After numerous perihelion passages, periodic comets finally lose all their volatile material. Only fragments of nonvolatile solids may survive. (Sections 11.8 through 11.10)

7. (a) 1; (b) 3; (c) 2. (Sections 11.12, 11.13, 11.16)

8. Meteor showers occur when Earth, moving along its orbit around the Sun, crosses a swarm of meteoroids left behind in space by a comet. (Section 11.14)

9. Iron meteorites—mostly iron (about 90 percent) and nickel; stony irons—iron, nickel, and silicates; stony meteorites—high silicate content, only about 10 percent iron and nickel by mass; probable origin: asteroid belt. (Section 11.17)

10. Earth, asteroid belt, Kuiper Belt, Oort comet cloud. (Sections 11.8, 11.17)

11. Because they originate in space and can help us understand the history and composition of our own planet, Earth, and of the solar system. (Sections 11.16, 11.17)

12

LIFE ON OTHER WORLDS?

We are attempting to survive our time so we may live into yours. We hope someday, having solved the problems we face, to join a community of galactic civilizations. This record represents our hope and determination, and our good will in a vast and awesome universe.

President Jimmy Carter, 1977
Record attached to Voyager spacecraft

Objectives

☆ Describe the molecular basis of Earth life.

☆ Give the evidence that indicates life may have evolved spontaneously from nonliving molecules on Earth.

☆ Explain a scientific theory of the origin and evolution of intelligent life on Earth.

☆ Describe the search for life on Mars.

☆ State the evidence for the existence of planetary systems other than our own.

☆ List the factors involved in estimates of the statistical chances for extraterrestrial intelligent life.

☆ Describe past and present human research and exploration in space.

☆ State the dominant current scientific view of interstellar voyages and UFOs.

☆ Describe several projects in which scientists have searched for or are planning to search for extraterrestrial intelligence.

12.1 PROMISE

No one knows whether **extraterrestrial life**, life beyond Earth, exists. Life on Earth could be a unique cosmic accident. But humans have compelling clues that we are not alone.

Biochemists find that all living organisms on Earth depend on water, a few basic organic molecules that they can manufacture from gas atoms in the laboratory, and energy.

Astronomers detect these basic atoms and molecules of life in our solar system, stars, and interstellar dust clouds. They have also found meteorites and comets with amino acids and water.

Physicists assume that the natural laws that rule physical and chemical events on Earth apply everywhere in the universe.

If life on Earth evolved from nonliving molecules in a series of physical and chemical processes, then life may have developed elsewhere among the more than 200 billion stars in our Milky Way Galaxy or in one of the 100 billion other galaxies in the universe.

The search has begun!

Astrobiology is the study of the origin, distribution, evolution, and future of life in the universe. Researchers at NASA's Astrobiology Institute test how life began and evolves on Earth and in space. ▶http://astrobiology.nasa.go/nai◀

Explain why many scientists think extraterrestrial life may exist. _____

_____

_____

_____

Answer: The basic molecules of life have been found in space and manufactured in the laboratory. If living organisms evolve from nonliving molecules in a series of physical and chemical reactions and are not a unique cosmic accident, then life may have occurred on other worlds.

12.2 COSMIC ORIGINS

The most basic characteristics that separate a living organism from a nonliving one are the ability to reproduce and metabolism. What causes the spark of life? No one knows, but a **cosmic evolution theory** ties the appearance of living organisms to universal forces in the following way:

Our universe began in the Big Bang 13.8 billion years ago. The first elements were hydrogen and helium. The universe expanded and cooled. Galaxies and stars formed. Heavier elements were slowly produced by nucleosynthesis inside massive stars. Supernovas produced the heaviest elements and sprayed enriched material back into space where it was recycled.

About 4.6 billion years ago, our Sun condensed from an interstellar cloud enriched with biotic elements and dust grains. Earth and the other solar system bodies took shape in a collapsing, cooling disk of material orbiting the infant Sun.

At first Earth's surface was tumultuous and fiery. Active volcanoes spewed hot lava and gases continually. Asteroids and comets crashed, adding more biotic elements to possibly seed life on the primitive Earth.

During the next billion years Earth cooled. Outgassing produced an atmosphere and ocean.

Scientists have energized a mixture of compounds of hydrogen, carbon, oxygen, and nitrogen gases like those in Earth's earliest atmosphere and have produced organic molecules, including amino acids, the basic molecules of life. The Sun's ultraviolet rays, cosmic rays, lighting flashes, and shock waves from geological activity were all available energy sources to bind the atmospheric gases into more complex organic molecules 4 billion years ago.

Another possibility is that hydrothermal vents on the ocean floor were the cradle of life, since we find organic molecules and life thriving in conditions there.

Gradually, organic molecules accumulated in Earth's seas and hot springs. As their concentration increased, collisions between molecules joined small molecules into larger ones. Water was important in this process, because it speeds chemical reactions by facilitating collisions between molecules.

Perhaps a billion years passed as more and more complex molecules formed. Eventually RNA (ribonucleic acid) and DNA (deoxyribonucleic acid) molecules were formed that carried the genetic codes for replication. The threshold from nonliving to living matter was crossed.

Why is water important in the chemical evolution of the basic organic molecules of life?_____

_____

Answer: Water speeds up chemical reactions by enabling molecules to collide with each other.

12.3 EVIDENCE

The common virus suggests that living organisms can evolve out of nonliving molecules, because it has characteristics of both.

A virus, too small to be seen other than under an electron microscope, is essentially a strand of DNA or RNA. A virus cannot provide its own energy or replicate itself outside of living cells. It survives when the cells it infects supply the energy necessary for growth and the means of reproduction.

If the sharp division between living and nonliving matter is artificial and viruses actually lie on a continuum, they would be somewhere near the middle. Perhaps a similar, ancient protocell was the precursor of life on Earth.

What evidence exists that living organisms may evolve out of nonliving molecules? _____

Answer: The virus has characteristics of both living organisms and nonliving molecules.

12.4 EVOLUTION

The principle of natural selection, or the survival of the fittest, asserts that all living creatures on Earth evolved from simple one-celled organisms.

Microfossils in rocks over 3 billion years old are evidence that life existed on Earth as simple one-celled plants called algae and organisms such as bacteria when those rocks formed (Figure 12.1).

Figure 12.1. Fossil bacteria, several billion years old.

The first living species reproduced. But offspring are never exact copies of their parents. Some variations in characteristics are always introduced whenever reproduction takes place.

Those individuals with favorable variations that helped them to survive the environment were the fittest. They had the best chance of reaching adulthood and reproducing themselves. Thus favorable traits were passed on, while unfavorable traits died out, by natural selection. Slowly, over a long period of time, a new species could have formed from the original one.

Multicelled organisms, which appeared about a billion years ago, and sexual reproduction speeded up evolutionary diversification.

The fossil record of the last 600 million years indicates that on many occasions there were mass extinctions of species followed by the arrival of numerous new, diverse species. The first fish were in the ocean about 425 million years ago. Reptiles appeared about 325 million years ago. After the extinction of the dinosaurs 65 million years ago, small mammals multiplied. Finally, some 40,000 years ago, humans with cognitive intelligence arrived.

In this way, over billions of years under the varying environmental conditions that existed on Earth, all living things, including modern humans, may have evolved from simple cells.

Suggest an environmental pressure that could produce a critical evolutionary advance._____

Answer: A drastic world-wide change in climate. (You may have thought of others.)

12.5 NEARBY PLANETS AND MOONS

Life may have evolved elsewhere in our solar system. The Sun's **habitable zone** (ecosphere), or region where life might most comfortably exist, is roughly between the orbits of Venus and Mars.

Venus does not seem suitable for life, because it has a prohibitively hot surface temperature up to 480° C (900° F) and is too dry.

Mars seems more suitable. It looks as though large amounts of water, essential for life on Earth, once flowed on Mars. Photographs show branching channels, valleys, and gullies that look as if they result from mighty floods, rivers, and water erosion. Water exists today in the Martian polar ice caps, frost, fog, and filmy clouds.

Extremophiles, microbes that thrive in environmental conditions of extreme temperature, pressure, or chemistry live on Earth. If ever life evolved on Mars, perhaps it still exists.

The U.S. Viking lander experiments were designed to detect carbon-based microbes living in the Martian soil. The results are inconclusive. Puzzling activity was detected. It might have been due to living organisms, or, more likely, to some unusual chemical characteristic of the Martian soil.

Jupiter and Titan may have simple microbes. Their clouds contain the gases out of which life probably evolved on Earth. There may be seas of liquid hydrogen on Jupiter or liquid methane on Titan where life could evolve. Europa, Callisto, Ganymede, and Enceladus may have life in subsurface oceans.

Scientists have not reached a consensus about the presence or absence of life on these other worlds. We must wait for planet probes of the future to find out.

Give three observations that suggest life may have developed from nonliving molecules on Mars. _____

_____

_____

Answer: (1) Evidence indicates that a lot of water once flowed on the planet. (2) Mars is inside the Sun's habitable zone. (3) Some plants and microbes can survive Martian conditions.

12.6 THE ODDS

We may be the only intelligent (as we proudly call ourselves) civilization in the entire universe. Or many others may exist.

Our Sun is just one of over 200 billion stars in our Milky Way Galaxy. Roughly a billion galaxies are within sight of our largest telescopes. Many of the stars in these galaxies may have life-supporting planets circling them. And some of these planets may bear intelligent civilizations.

A way to estimate the number of intelligent civilizations in our own Milky Way Galaxy—the only ones we could hope to communicate with at present—was suggested by astronomers Carl Sagan (1934–1996) and Frank Drake of the U.S. and I. S. Shklovsky (1916–1985) of Russia. First, estimate with some confidence (1) the number of stars in the Galaxy, (2) the fraction of those stars that have planets circling them, and (3) the average number of planets suitable for life.

With much less confidence, estimate (4) the fraction of suitable planets on which life actually developed, (5) the fraction of those life starts that evolved to intelligent organisms, and (6) the fraction of intelligent species that have attempted communication.

Then, just guess (7) the average lifetime of an intelligent civilization.

When all these factors are considered, estimates range anywhere from one intelligent civilization (ours) to a million in the Milky Way Galaxy today.

Why do you think (7), the average lifetime of an intelligent civilization, is the most uncertain number of all? _____

_____

_____

_____

_____

Answer: No one knows what happens when a civilization like ours, if we are typical, reaches a stage of technical development where it can communicate with other civilizations in our Galaxy. Will it last long enough for a conversation, or will it self-destruct with nuclear weapons, pollution, or overpopulation?

12.7 EXTRASOLAR PLANETARY SYSTEMS

The nebular theory of the formation of stars asserts and new observations confirm that many stars have planets (Section 4.3).

Circumstellar disks, swarms of gases and particles orbiting stars, were first observed at infrared wavelengths in 1983 (Figure 12.2). Direct photo-

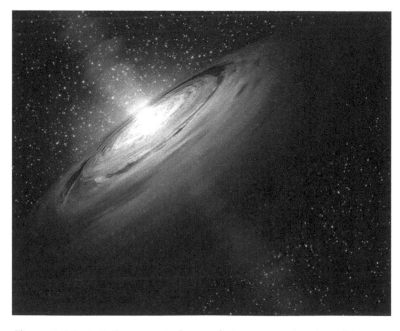

Figure 12.2. Artist's concept of a proplyd, or protoplanetary disk.

graphic confirmation came first for nearby Beta Pictoris, 50 light-years away. Thicker disks around young stars are likely planetary systems forming, called proplyds or protoplanetary disks. Thinner disks around older stars are leftovers from planets that have already formed.

Extrasolar planets, or **exoplanets**, belonging to stars other than the Sun, are hard to see. Astronomers commonly search for unseen companions using indirect techniques:

1. Astrometry: Measure the proper motion of the candidate visible star.

The gravitational tug of planets would cause a tiny wobble in the otherwise straight-line path of a star across the sky. Barnard's star in Ophiuchus, with the largest proper motion, was photographed for years. A wobble only 0.01 as big as the star's image was reported and attributed to planets. None has been confirmed.

2. Spectroscopy: Measure the radial velocity of the candidate visible star.

The gravitational tug of planets causes a small but measurable periodic Doppler shift in a star's spectral lines. In 1995 Sun-like star 51 Pegasi, 50 light-years away, showed slight radial velocity variations, caused by the first confirmed extrasolar planet. It is a hot, half-Jupiter-size planet orbiting close to its star every 4.2 days.

Many more such discoveries quickly followed. Hot, giant planets close to their stars are easiest to find. They tug harder and orbit quicker than lower mass and more distant planets.

In 1999 the first extrasolar planetary system was detected. Three massive planets orbit Upsilon Andromedae, an F-type star that is 44 light-years away.

3. Photometry: Measure the light output of the candidate visible star.

A planet that moves directly in front of its star along our line of sight causes a slight dip in brightness. Many stars must be surveyed because orbits must be favorably inclined to us. Starlight from HD 209458 in Pegasus dimmed slightly in 1999. The dip was attributed to the first planet observed by its transit.

U.S. robot Kepler (2009–) found thousands of possible gas giants, hot-super-Earths in short-period orbits, and ice giants this way. Its goal is to find Earth-like planets and to figure how many stars in our Galaxy have such planets. ▶kepler.nasa.gov◀

4. Gravitational microlensing: Measure light amplification by the candidate visible star.

If a faint star is along our line of sight to a brighter, more distant star, its gravitational field acts as a lens to amplify the bright starlight. Planets orbiting the intervening star cause tiny changes in this light amplification.

5. Direct Imaging: Take a picture of the exoplanet in visible light.

An exoplanet was first imaged in 2008. This method works best for large, hot, young planets far from their stars that emit infrared light. A stellar coronagraph blocks the star's bright light.

The hunt for more giant and Earth-like planets proceeds with ever improving ground- and space-based tools. Ultimately we may see an extrasolar planet directly! Keep up with planet discoveries at the Extrasolar Planets Encyclopedia. ▶Planetquest.jpl.nasa.gov◀

What does a wobble in the motion of a visible star suggest?

___

Answer: The star's wobble suggests the presence of unseen companions.

12.8 SPACE TRAVEL

Interstellar travel, trips to other stars, would be the most dramatic way to see if other civilizations exist. But we are not yet ready for a deep space voyage.

Even the closest stars are several light-years away, and none of our spacecraft can travel anywhere near the speed of light. A trip to our Sun's nearest bright neighbor, Alpha Centauri, at the speed the Apollo astronauts traveled to the Moon, would require thousands of years.

Russian cosmonauts were the first humans in space. Yuri Gagarin (1934–1968) orbited Earth once in Vostok 1 and landed after 1 hour, 48 minutes on April 12, 1961. Valentina V. Tereshkova, the first female, circled Earth 48 times June 16–19, 1963.

Figure 12.3. U.S. piloted spacecraft (1961–2011). Spacecraft / rocket launchers, left to right: Mercury / Atlas; Gemini / Titan 2; Apollo / Saturn 5; Skylab / Saturn 5; Apollo / Saturn 1B–Soyuz; Space Shuttle Orbiter / External fuel tank and solid rocket boosters.

For the next 20 years Russia and the U.S. developed increasingly sophisticated piloted spacecraft. Each flew just once, and most are on display in space museums today (Figure 12.3).

Astronauts now do microgravity experiments aboard space stations, Earth-orbiting satellites worked by rotating crews. From 1981 to 2011 they also flew satellites to and from space and made repairs aboard space shuttles, reusable craft whose missions lasted a few weeks in Earth orbit (Figure 12.4).

Long periods of weightlessness change human physiology, body chemistry, and mental health measurably. Russian cosmonaut Valery Polyaikov set the world record for time in space in 1995: 438 days, 18 hours.

Today the U.S., Canada, Japan, Russia, 11 European Space Agency nations, and Brazil fly the International Space Station (ISS). Current research focuses on the biomedical effects of microgravity and on ways to help people adapt to it, as well as on the exploration and utilization of space (Figure 12.5).

When did the first human go into space? _____

Answer: 1961 (Yuri Gagarin, April 12, 1961).

Figure 12.4. Space shuttles flew modules to the International Space Station, where construction was completed. Astronaut Rick Linnehan is anchored to a mobile foot restraint as he works outside during a seven-hour spacewalk.

Figure 12.5. The 924,000-pound International Space Station in Earth orbit. The length and width of a football field, it has two bathrooms, a gym, a 360-degree bay window, and more livable room than a six-bedroom house. Eight solar panels supply electricity.

12.9 STAR PROBES

At present, to send high-speed robot probes to other stars would be prohibitively expensive.

Four U.S. spacecraft are escaping into interstellar space after having completed missions to the giant planets. They carry symbolic messages for any intelligent beings they may encounter far beyond our solar system.

Pioneer 10, the first spacecraft to penetrate the asteroid belt and to provide closeups of Jupiter in 1973, was the first to fly beyond Neptune's orbit, in 1983. Its twin, Pioneer 11, followed in 1990. Each bears a symbolic plaque that tells about us (Figure 12.6). At last contact on January 22, 2003, Pioneer 10 was 12.2 billion km (7.6 billion miles) from Earth.

Voyager 1 and 2 (Section 8.12) carry a unique record with electronically encoded information, sounds, and pictures of the best of Earth and its inhabitants. A cartridge and playing instructions are included. Possible civilizations in space can hear recordings of wind and waves, birds and animals, music, a kiss and a baby's cry, and greetings in 60 languages.

Voyagers are exploring the region at the edge of our solar system. In 2013 Voyager 1's fields and particles instruments reported the heliopause, where

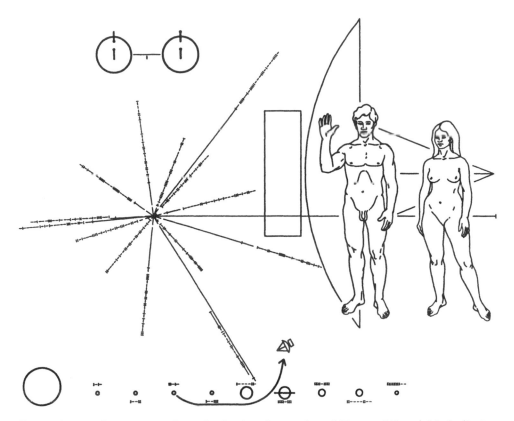

Figure 12.6. First message from Earth, the plaque aboard Pioneer 10 and 11, indicates when, where, and by whom each spacecraft was launched.

the Sun's influence ends and interstellar space begins. The Voyagers should return valuable data until the year 2020 when their nuclear power sources will no longer be able to supply necessary electrical energy.

People sometimes wonder whether Earth is visited by creatures from other worlds when they hear accounts of unidentified flying objects (UFOs).

Most scientists think aliens are the least likely explanation for UFO sightings. They ask for concrete evidence, such as a piece of an alien spacecraft, to examine in a laboratory. None has been found so far.

Why is it unlikely that Earth will be invaded by hostile beings from space, as some alarmists have suggested? _____

_____

_____

_____

Answer: Travel across the enormous distances between stars would take much too long and exhaust too many resources to be justifiable for most purposes, if other civilizations are at our stage of development.

12.10 COMMUNICATION

We have the capability of communicating at the speed of light with other civilizations using radio waves.

With sending and receiving devices already in use, we can radio a message from Earth that could be detected by another civilization like ours across the Galaxy. We are capable of detecting radio messages from radio telescopes located thousands of light-years away but are no more powerful than ours.

One coded message from Earth was radioed into space in 1974, largely to show off the capabilities of the giant radio telescope at Arecibo, Puerto Rico. This signal was beamed toward the globular cluster M13 in the constellation Hercules, 24,000 light-years away. The least amount of time required to get an answer from M13 at the speed of light is 48,000 years!

Current research concentrates on receiving intelligent radio signals from other civilizations beyond our solar system. This is cheaper, simpler, and safer than deliberately sending out signals until we know where other civilizations are and if they are friendly.

The first attempt to receive intelligent signals was Project Ozma at the U.S. National Radio Astronomy Observatory in Green Bank, West Virginia. Astronomer Frank Drake listened at a frequency of 1420 MHz (21 cm wavelength) to two nearby stars, Tau Ceti and Epsilon Eridani. No intelligent signals were detected then or subsequently.

This lack of success is not at all surprising. Even if other civilizations were deliberately trying to make themselves known to us, we might not have looked in the right direction at the right time, or tuned to the right frequency. The process is rather like trying to find a needle in a haystack by looking only occasionally, and without even knowing what a needle looks like.

About how long would it take a radio-wave message from Earth to reach the Sun's closest neighbor star system, Alpha Centauri (4.4 light-years away)? _____

Answer: 4.4 years. (Radio waves travel at the speed of light.)

12.11 A SERIOUS SEARCH

Scientists can search for extraterrestrial intelligence using existing radio telescopes and computers, referred to as SETI projects.

Figure 12.7. The Allen Telescope Array (ATA) in northern California is the most sophisticated SETI project ever. Some of its planned 350 radio dishes are working together with a wide field of view to detect modulated signals produced by intelligent aliens. The ATA conducts both all-sky and targeted searches 24/7.

The challenge is to search for a weak, undefined radio signal from an unknown direction, since no one knows exactly how far away or where possible alien transmitters are or which frequencies they use (Figure 12.7).

The frequencies considered most likely for our first contact are between 1400 MHz and 1700 MHz, often referred to as the galactic "waterhole" where we shall all meet. A modulated signal in the 21-cm microwave region would stand out because astronomers anywhere in our Galaxy would recognize it. Also, a transmitter would require the least power there to generate a detectable signal above the natural background noise (Section 6.8).

Hopes rest on affordable automated systems in which computers act as sensitive multichannel spectrum analyzers (MCSA), or radio receivers for a wider band of frequencies that scan millions of radio channels at once.

Current searches for extraterrestrial civilizations include two complementary types of strategies:

1. An **all-sky survey** that searches the entire sky over a wide frequency range to detect strong signals. Telescopes in the northern and southern hemispheres scan over frequencies from 1000 MHz to 10,000 MHz and some accessible frequencies to 25,000 MHz.

2. A high-sensitivity **targeted search** that seeks weak signals originating near close stars like our Sun. This targets stars within 100 light-years of Earth over the frequency range 1000 MHz to 3000 MHz plus any accessible frequencies up to 10,000 MHz.

In the future, we can construct even more sensitive multichannel analyzers and bigger antennas or antenna arrays. Alternatives such as putting antennas in space or on the Moon are also possibilities.

Imagine that as a citizen you have been asked to cast a vote for or against our joining a serious, affordable international search for signals from other civilizations. How would you vote, and why? _____

_____

_____

_____

_____

Answer: Your answer is a matter of opinion. I would vote for it. If we found intelligent aliens, they might teach us how to surmount problems now threatening our survival on Earth. If not, the money would still be well spent. We could expect tremendous gains in peaceful coexistence and knowledge for humanity's benefit from an international commitment to an intellectual and scientific effort of this size.

 Join the search for signals from an alien civilization. ▶SETILIVE.org◀

SELF-TEST

This self-test is designed to show you whether or not you have mastered the material in Chapter 12. Answer each question to the best of your ability. Correct answers and review instructions are given at the end of the test.

1. List two observations that support the theory that the first living things on Earth may have evolved spontaneously out of nonliving chemicals.

   (1)_____

   _____

   _____;

   (2)_____

   _____

2. Briefly summarize the scientific theory of the evolution of intelligent life on Earth from simple one-celled organisms._____

   _____

   _____

   _____

   _____

   _____

   _____

   _____

   _____

3. Explain why the first search for life on another planet was conducted on Mars. _____

   _____

   _____

   _____

   _____

4. Describe two types of observations that give evidence for planets circling stars beyond our Sun. (1) _____

_____

_____

(2)_____

_____

_____

_____

5. What is the most uncertain factor involved in estimates of the statistical chances for extraterrestrial intelligent life?_____

_____

6. What is the dominant current scientific view of UFOs? _____

_____

_____

_____

7. Match the following firsts to the correct spacecraft. First:

_____ (a) Human in space (Yuri Gagarin).   (1) Space shuttle.

_____ (b) Reusable piloted spacecraft.   (2) Pioneer 10.

_____ (c) Spacecraft to cross Neptune's orbit.   (3) Viking 1 and 2.

_____ (d) Search for life on surface of   (4) Vostok 1.
           another planet.   (5) Voyager 1.

_____ (e) Cross the heliopause into interstellar space.

8. Explain why scientists concentrate on receiving intelligent radio signals in their search for possible life on planets circling other stars beyond our Sun.

_____

_____

_____

_____

9. List two complementary search strategies that are popular today.

(1)_____

_____

_____

_____ ;

(2)_____

_____

_____

ANSWERS

Compare your answers to the questions on the self-test with the answers given below. If you missed any questions, review the sections indicated in parentheses following the answer. If you missed several questions, you should probably reread the entire chapter carefully. Look in the Appendixes for other interesting and useful information about astronomy.

1. (1) Biologists find that all living organisms on Earth depend on water and a few basic organic molecules that have been manufactured by energizing gas atoms in laboratories. (2) The common virus has characteristics of both living organisms and nonliving molecules.   (Sections 12.1, 12.3)

2. The principle of natural selection, or the survival of the fittest, asserts that when living species reproduced, variations in characteristics were introduced. Those individuals with favorable variations that helped them to survive the environment were the fittest. They had the best chance of reaching adulthood and reproducing themselves. Thus favorable traits were passed on while unfavorable traits died out, by natural selection. Intelligence is a favorable trait. Over a period of millions of years under the different environmental conditions that existed on Earth, intelligent beings evolved from simple cells.   (Section 12.4)

3. Mars is in the Sun's habitable zone. Evidence suggests that water once flowed on Mars. Certain terrestrial plants and microbes can survive environmental conditions similar to those on Mars. Mars is close enough for the trip to be cost effective.   (Sections 12.5, 12.8, 12.9)

4. (1) Direct observation: A visible light or infrared image; a circumstellar disk, which could be a planetary system in its early formative stage. (2) Indirect observation: A perturbation that could be caused by the gravitational tug of massive planets, such as a wobble in the proper motion of a star; a Doppler shift in the spectral lines of a star, which indicates radial velocity changes; or a slight dip in brightness of a visible star that could occur when a planet moves directly in front of it.   (Section 12.7)

5. The average lifetime of an intelligent civilization.   (Section 12.6)

6. Most scientists do not think we have been visited by creatures from other worlds. They would be glad to examine concrete evidence, such as a piece of an alien spacecraft, in a laboratory, but none has been found.   (Section 12.9)

7. (a) 4;   (b) 1;   (c) 2;   (d) 3;   (e) 5.   (Sections 12.1, 12.8, 12.9)

8. Even the closest stars are several light-years away, so we cannot travel there. We have the capability of communicating at the speed of light with other civilizations using radio waves. Concentrating on receiving intelligent signals

is cheaper, simpler, and safer than transmitting until we know where other civilizations are and if they are friendly.   (Sections 12.8 through 12.10)

9. (1) All-sky survey that searches the entire sky over a wide frequency range to detect strong signals. (2) High-sensitivity targeted search directed at nearby stars like our Sun, over a narrower frequency range.   (Section 12.11)

EPILOGUE

*Astronomy compels the soul to look upwards
and leads off from this world to another.*

Plato (c. 428–348 B.C.)
The Republic

Astronomy has come a long way since the ancients first pondered the mysteries of the universe. But many exciting discoveries still lie ahead. Now that you have deepened your understanding of basic ideas, you should enjoy observing the sky and contemporary discoveries more than ever!

USEFUL RESOURCES:
PRINT AND ONLINE

PERIODICALS

Air and Space—Smithsonian ▶www.airspacemag.com◀

Astronomy ▶www.astronomy.com◀

Discover ▶discovermagazine.com◀

Griffith Observer ▶www.griffithobs.org/observer.html◀

Mercury ▶www.astrosociety.org◀

National Geographic ▶www.nationalgeographic.com◀

Natural History ▶www.naturalhistorymag.com◀

Science News ▶www.sciencenews.org◀

Sky and Telescope ▶www.skyandtelescope.com◀

DATABASES

The author found the following databases useful for her research work.

NASA/JPL Solar System Dynamics ▶http://ssd.jpl.nasa.gov◀
Orbits, physical characteristics, and discovery information for most known natural bodies in orbit around our Sun.

SIMBAD ▶http://simbad.u-strasbg.fr◀
Database operated at Centre de Données Astronomiques de Strasbourg. Basic data, cross-identifications, bibliography, and measurements for objects outside the solar system.

NASA/IPAC ExtraGalactic Database ▶http://nedwww.ipac.caltech.edu◀
Comprehensive multi-wavelength database of fundamental measurements for known objects beyond the Milky Way.

NASA/NSSDC Space science mission data. ▶nssdc.gsfc.nasa.gov◀
Permanent archive for astronomy and astrophysics, solar and space plasma physics, space physics, and planetary and lunar science mission data.

BOOKS BY DINAH L. MOCHÉ, Ph.D.

▶www.spacelady.com◀

AMAZING ROCKETS, Golden Books, New York, NY

AMAZING SPACE FACTS, 2nd edition, Golden Books, New York, NY

ASTRONOMY, 7th edition, John Wiley & Sons, Hoboken, NJ

ASTRONOMY TODAY, 2nd edition, Random House, New York, NY

THE ASTRONAUTS, Random House, New York, NY

THE GOLDEN BOOK OF SPACE EXPLORATION, Golden Books, New York, NY

IF YOU WERE AN ASTRONAUT, 2nd edition, Golden Books, New York, NY

LABORATORY MANUAL FOR INTRODUCTORY ASTRONOMY—founding editor and co-author

LIFE IN SPACE, Ridge Press / A & W Publishers, New York, NY

MAGIC SCIENCE TRICKS, Scholastic Book Services, New York, NY

MARS, Franklin Watts, New York, NY

MORE MAGIC SCIENCE TRICKS, Scholastic Book Services, New York, NY

MY FIRST BOOK ABOUT SPACE, Golden Books, New York, NY

RADIATION, Franklin Watts, New York, NY

SEARCH FOR LIFE BEYOND EARTH, Franklin Watts, New York, NY

WHAT'S UP THERE? QUESTIONS AND ANSWERS ABOUT STARS AND SPACE, updated regularly, Scholastic Book Services, New York, NY

WHAT'S DOWN THERE? QUESTIONS AND ANSWERS ABOUT THE OCEAN, Scholastic Book Services, New York, NY

WE'RE TAKING AN AIRPLANE TRIP, Golden Books, New York, NY

CAREER INFORMATION

A Career in Astronomy, The American Astronomical Society, Director of Educational Activities, AAS Executive Office, 2000 Florida Avenue, NW, Suite 400, Washington, DC 20009-1231, ▶www.aas.org/education/publications/careerbrochure.pdf◀

Degree Programs in Physics and Astronomy in U.S. Colleges and Universities, American Institute of Physics, ▶www.aip.org◀

(a) Physics in Your Future, 2nd edition, free from American Physical Society, ▶www.aps.org/programs/women/reports/physicsfuture/upload/piyf.pdf◀, and *(b) Women in Science* by Dinah L. Moché, Ph.D., American Association of Physics Teachers, One Physics Ellipse, College Park, MD 20740. (a) Career information for junior and senior high school, and (b) a multimedia package presenting information about six women scientists and their work.

ALMANACS, OBSERVING GUIDES, AND STAR ATLASES

A Field Guide to the Stars and Planets, 4th edition, by Wil Tirion (Illustrator) and Jay M. Pasachoff (Boston: Houghton Mifflin Co., 1999).

The Astronomical Almanac, and *The Astronomical Almanac* Online issued annually by the U.S. Naval Observatory (Washington, DC: U.S. Government Printing Office, ▶http://www.gpo.gov◀, yearly). Current information about Sun, Moon, planets, eclipses, and occultations.

Burnham's Celestial Handbook, Volumes 1, 2, and 3, Revised edition (New York: Dover Publications, Inc., 1980). Observer's guide to space beyond the solar system.

Observer's Handbook, edited by Dave Chapman, issued annually by The Royal Astronomical Society of Canada, 203-4920 Dundas Street West, Toronto, Ontario M9A 1B7. Information and tables on the Sun, Moon, planets, asteroids, meteor showers, and other celestial phenomena.

Sky Atlas 2000.0, 2nd edition, by Wil Tirion (Cambridge, MA: Cambridge University Press, 1999).

Sky Calendar (East Lansing, MI: Abrams Planetarium, Michigan State University, yearly).

ORGANIZATIONS

American Association of Variable Star Observers
49 Bay State Rd
Cambridge, MA 02138
(617) 354-0484
▶www.aavso.org◀

American Association of Physics Teachers
One Physics Ellipse
College Park, MD 20740
(301) 209-3333
▶www.aapt.org◀

American Astronomical Society
2000 Florida Avenue, NW, Suite 400
Washington, DC 20009-1231
(202) 328-2010
▶www.aas.org◀

Astronomical League National Headquarters
9201 Ward Parkway, Suite 100
Kansas City, MO 64114
(816) 333-7759
▶www.astroleague.org◀

Astronomical Society of the Pacific
390 Ashton Avenue
San Francisco, CA 94112
(415) 337-1100
▶www.astrosociety.org◀

British Astronomical Association
Burlington House, Piccadilly
London W1J ODU, UK
▶britastro.org◀

International Astronomical Union
IAU - UAI Secretariat
98-bis Blvd Arago
F–75014 PARIS
FRANCE

►www.iau.org◄

International Dark-Sky Association
3223 N. First Avenue
Tucson, AZ 85719
(520) 293-3198

►www.darksky.org◄

International Planetarium Society
c/o Shawn Laatsch, Treasurer
600 Imiloa Place
Hilo, HI 96720

►www.ips-planetarium.org◄

National Science Teachers Association
1840 Wilson Blvd.
Arlington, VA 22201
(703) 243-7100

►www.nsta.org◄

National Space Society
1155 15th Street NW, STE 500
Washington, DC 20003
(202) 429-1600

►www.nss.org◄

Royal Astronomical Society
Burlington House, Piccadilly
London W1J OBQ, UK

►www.ras.org.uk◄

Royal Astronomical Society of Canada
203-4920 Dundas St W
Toronto, Ontario, Canada, M9A 1B7
(416) 924-7973

►www.rasc.ca◄

The Planetary Society
85 South Grand Avenue
Pasadena, CA 91105
(626) 793-5100

►http://planetary.org◄

STUNNING COLOR IMAGES AND NEWS ONLINE

The quantity of spectacular images and the new information about astronomy and space exploration online increases daily. Besides the websites given in the text, those following give you quick access to the best astronomy and space sites on the Internet.

Spectacular Images

Anglo-Australian Observatory Image Collection
New photographic and CCD images of stars, galaxies, and nebulas from Australia.
►www.aao.gov.au/images◄

Astronomy Picture of the Day
A different image every day, with explanations, links, and archive.
▶antwrp.gsfc.nasa.gov/apod/astropix.html◀

Hubble Heritage Photos
Greatest hits from Hubble Space Telescope, added monthly and archived.
▶heritage.stsci.edu/◀

NASA Images
Comprehensive compilation of NASA's vast collection of photographs, film, and video.
▶www.nasaimages.org◀

NASA's Planetary Photo Journal
Publicly released images from Solar System exploration programs.
▶photojournal.jpl.nasa.gov/◀

National Optical Astronomy Observatories Image Gallery
Images from Kitt Peak, NSO/Sac Peak, and Gemini telescopes.
▶www.noao.edu/◀

Space Telescope Science Institute
Hubble Space Telescope pictures and operation.
▶www.stsci.edu◀

General Astronomy and Space Exploration

CAPjournal
Free peer-reviewed journal for astronomy communicators online and in print, from the International Astronomical Union.
▶www.iau.org◀

ESA
News, photos, and videos from the European Space Agency's piloted and robotic space missions.
▶www.esa.int◀

Google Sky
Images, maps, and information. Also online: Google Earth/Moon/Mars
▶www.google.com/sky◀

Harvard-Smithsonian Center for Astrophysics
Recommended astronomy apps for iPhone and Android, *Stargazing with your Smart Phone and Tablet.*
▶*http://www.youtube.com/watch?v=b05GAQa8GeQ*◀

NASA Science
In-depth coverage of NASA's past, present, and future science missions; NASA apps for mobile devices; podcasts; for researchers, educators, kids, and citizen scientists.
▶www.nasa.gov◀

JAXA

Space policies, activities, news, and images by Japan's national aerospace agency, Japan Aerospace Exploration Agency.
►http://global.jaxa.jp◄

National Air and Space Museum, Smithsonian Institution

Largest collection of historic air- and spacecraft in the world.
►www.nasm.si.edu/◄

Portal to the Universe

Aggregation of news, images, events, videos, educational materials, addresses for facilities and societies, and a social network.
►www.portaltotheuniverse.org/◄

World Wide Telescope (WWT) from Microsoft

Imagery from the best ground- and space-based observatories across the world. Allows users to explore the night sky through their computers.
►www.worldwidetelescope.org◄

Teaching Materials

American Astronomical Society Education Office

Resources for undergraduate and graduate teachers, K–12, and public outreach.
►www.aas.org/education/◄

Center for Astronomy Education

Resources and useful links for the professional development of introductory astronomy instructors.
►http://astronomy101.jpl.nasa.gov◄

Galileoscope

A high-quality, low-cost telescope kit developed by a team of leading astronomers, optical engineers, and science teachers. Easy-to-assemble. Educational and observational resources for educators available.
►http://galileoscope.org◄

NASA Education Program

NASA resources, programs, multimedia products, and calenders for educators, students, and the informal education community.
►http://education.nasa.gov◄

National Optical Observatories Educational Outreach Program

Information, programs, and materials for K–12 and college by NOAO scientists.
►www.noao.edu/education/noaoeo.html◄

Sky Publishing Astronomy Education Links

Organizations, media, general astronomy references, solar system, teaching resources, courses, and camps.
►www.skyandtelescope.com/◄

Appendix 1
THE CONSTELLATIONS

Name and Pronunciation	Genitive	Abbr.	Meaning	RA	Decl.
Andromeda, ăn-drŏm′ é-dà	-dae	And	Daughter of Cassiopeia	1^h	+40°
Antlia, ănt lĭ-à	-liae	Ant	The Air Pump	10^h	−35°
Apus, ā pŭs	Apodis	Aps	Bird of Paradise	16^h	−75°
Aquarius, à-kwâr ĭ-ŭs	-rii	Aqr	The Water-bearer	23^h	−10°
Aquila, ăk wĭ-là	-ae	Aql	The Eagle	19^h30^m	+ 5°
Ara, ā rà	Arae	Ara	The Altar	17^h30^m	−55°
Aries, ā rĭ-ēz	Arietis	Ari	The Ram	2^h	+20°
Auriga, ô-rī′ gà	-gae	Aur	The Charioteer	5^h30^m	+40°
Bootes, bō-ō tēz	-tis	Boo	The Herdsman	14^h30^m	+30°
Caelum, sē lŭm	Caeli	Cae	The Graving Tool	4^h30^m	−40°
Camelopardalis kà-měl′ ō-pàr′ dà-lĭs	Camelopardalis	Cam	The Giraffe	6^h	+70°
Cancer, kăn′ sēr	Cancri	Cnc	The Crab	8^h30^m	+20°
Canes Venatici kā′ nēz vē-năt′ ĭ-sī	Canum Venati-corum	CVn	The Hunting Dogs	12^h30^m	+40°
Canis Major, kā′ nĭs mā′ jēr	Canis Majoris	CMa	The Big Dog	7^h	−20°
Canis Minor, kā′ nĭs mī′ nēr	-ris	CMi	The Little Dog	7^h30^m	+ 5°
Capricornus, kăp′ rĭ-kôr′ nŭs	Capricorni	Cap	The Horned Goat	21^h	−20°
Carina, kà-rī′ nà	-nae	Car	The Ship's Keel	9^h	−60°
Cassiopeia, kăs′ ĭ-ō-pē′ yà	Cassiopeiae	Cas	The Queen	1^h	+60°
Centaurus, sěn-tô′ rŭs	-ri	Cen	The Centaur	13^h	−50°
Cepheus, sē′ fŭs	Cephei	Cep	The King	22^h	+65°
Cetus, sē′ tŭs	Ceti	Cet	The Whale	2^h	−10°
Chamaeleon, kà-mē′ lē-ŭn	-ntis	Cha	The Chameleon	10^h	−80°
Circinus, sûr′ sĭ-nŭs	-ni	Cir	The Compasses	15^h	−60°
Columba, kō-lŭm′ bà	-bae	Col	The Dove	6^h	−35°
Coma Berenices kō′ mà běr′ ě-nī sēz	Comae Berenices	Com	Berenice's Hair	13^h	+25°
Corona Australis kō-rō′ nà ôs-trā′ lĭs	-nae	CrA	The Southern Crown	19^h	−40°
Corona Borealis kō-rō′ nà bō′ rē-ā′ lĭs	Coronae Borealis	CrB	The Northern Crown	15^h30^m	+30°
Corvus, kôr′ vŭs	Corvi	Crv	The Crow	12^h30^m	−20°
Crater, krā′ tēr	-ris	Crt	The Cup	11^h30^m	−15°
Crux, krŭks	Crucis	Cru	The Southern Cross	12^h30^m	−60°
Cygnus, sĭg′ nŭs	Cygni	Cyg	The Swan	20^h	+40°
Delphinus, děl-fī′ nŭs	-ni	Del	The Dolphin	20^h30^m	+15°
Dorado, dō-rà′ dō	-dus	Dor	The Swordfish	5^h	−60°
Draco, drā′ kō	Draconis	Dra	The Dragon	18^h	+70°
Equuleus, ē-kwoō′ lē-ŭs	-lei	Equ	The Colt	21^h	+10°
Eridanus, ē-rĭd′ à-nŭs	Eridani	Eri	The River	3^h30^m	−20°
Fornax, fôr′ năks	-nacis	For	The Furnace	3^h	−30°
Gemini, jěm′ ĭ-nī	Germinorum	Gem	The Twins	7^h	+25°
Grus, grŭs	Gruis	Gru	The Crane	22^h	−40°
Hercules, hûr′ kū-lēz	Herculis	Her	Hercules	17^h	+35°
Horologium, hŏr′ ō-lō′ jĭ-ŭm	-gii	Hor	The Clock	3^h	−50°
Hydra, hī′ drà	Hydrae	Hya	The Water Snake (female)	11^h	−20°
Hydrus, hī′ drŭs	-dri	Hyi	The Water Snake (male)	2^h	−70°
Indus, ĭn′ dŭs	Indi	Ind	The Indian	21^h	−50°
Lacerta, la-sûr′ tà	-tae	Lac	The Lizard	22^h30^m	+50°
Leo, lě′ ō	Leonis	Leo	The Lion	10^h30^m	+20°
Leo Minor, lě′ ō mī′ nēr	Leonis Minoris	LMi	The Little Lion	10^h30^m	+35°
Lepus, lě′ pŭs	Leporis	Lep	The Hare	5^h30^m	−20°

(continued)

341

Name and Pronunciation	Genitive	Abbr.	Meaning	Reference RA	Decl.
Libra, lī′ brȧ	Librae	Lib	The Balance	15^h	−15°
Lupus, lū′ pŭs	Lupi	Lup	The Wolf	15^{h}30^m	−40°
Lynx, lĭnks	Lyncis	Lyn	The Lynx	8^h	+50°
Lyra, lī′ rȧ	Lyrae	Lyr	The Lyre	18^{h}30^m	+35°
Mensa, mĕn′ sȧ	-sae	Men	Table Mountain	5^{h}30^m	−75°
Microscopium mī′ krŏ-skō′ pĭ-ŭm	-pii	Mic	The Microscope	21^h	−35°
Monoceros, mō-nŏs′ ĕr-ŏs	Monocerotis	Mon	The Unicorn	7^h	− 5°
Musca, mŭs′ kȧ	-cae	Mus	The Fly	12^h	−70°
Norma, nôr′ ma	-mae	Nor	The Square	16^h	−50°
Octans, ŏk′ tănz	-ntis	Oct	The Octant	0^h-25^h	−90°
Ophiuchus, ŏf′ ĭ-ū′ kŭs	Ophiuchi	Oph	The Serpent-bearer	17^h	0°
Orion, ō-rī′ ŏn	Orionis	Ori	The Hunter	5^{h}30^m	0°
Pavo, pā′ vō	-vonis	Pav	The Peacock	20^h	−65°
Pegasus, pĕg′ ȧ-sŭs	Pegasi	Peg	The Winged Horse	23^{h}30^m	+20°
Perseus, pûr′ sūs	Persei	Per	Perseus	3^{h}30^m	+45°
Phoenix, fē′ nĭks	Phoenicis	Phe	The Phoenix	1^h	−50°
Pictor, pĭk′ tēr	Pictoris	Pic	The Painter's Easel	6^h	−55°
Pisces, pīs′ ēz	Piscium	Psc	The Fishes	23^{h}30^m	+ 5°
Piscis Austrinus pīs′ is ôs-trī′ nŭs	Piscis Austrini	PsA	The Southern Fish	23^h	−30°
Puppis, pŭp′ ĭs	Puppis	Pup	The Ship's Stern	8^h	−40°
Pyxis, pĭk′ sĭs	Pyxidis	Pyx	The Compass	9^h	−30°
Reticulum, rē-tĭk′ ū-lŭm	-li	Ret	The Net	4^h	−60°
Sagitta, sȧ-jĭt′ ȧ	-tae	Sge	The Arrow	20^h	+20°
Sagittarius, săj′ ĭ-tā′ rĭ-ŭs	Sagittarii	Sgr	The Archer	18^{h}30^m	−30°
Scorpius, skôr′ pĭ-ŭs	Scorpii	Sco	The Scorpion	17^h	−30°
Sculptor, skŭlp′ tēr	-ris	Scl	The Sculptor	0^h	−30°
Scutum, skū′ tŭm	Scuti	Sct	The Shield	18^{h}30^m	−10°
Serpens, sûr′ pĕnz	Serpentis	Ser	The Serpent	16^h	0°
Sextans, sĕks′ tănz	-ntis	Sex	The Sextant	10^h	− 5°
Taurus, tô′ rŭs	Tauri	Tau	The Bull	4^{h}30^m	+15°
Telescopium, tĕl′ ē-skō′ pĭ-ŭm	-pii	Tel	The Telescope	19^h	−50°
Triangulum, trī-ăng′ gū-lŭm	Trianguli	Tri	The Triangle	2^h	+30°
Triangulum Australe trī-ăng′ gū-lŭm ôs-trā′ lē	Trianguli Australis	TrA	The Southern Triangle	16^h	−65°
Tucana, tū-kā′ nȧ	-nae	Tuc	The Toucan	23^{h}30^m	−65°
Ursa Major, ûr′ sȧ mā′ jēr	Ursae Majoris	UMa	The Great Bear	11^h	+60°
Ursa Minor, ûr′ sȧ mī′ nēr	Ursae Minoris	UMi	The Little Bear	15^h	+70°
Vela, vē′ lȧ	Velorum	Vel	The Ship's Sails	9^h	−50°
Virgo, vûr′ gō	Virginis	Vir	The Maiden	13^h	−10°
Volans, vō′ lănz	-ntis	Vol	The Flying Fish	8^h	−70°
Vulpecula, vŭl-pĕk′ ū-lȧ	Vulpeculae	Vul	The Fox	19^{h}30^m	+25°

ā dāte; ă tăp; â câre; ȧ ȧsk; ē wē; ĕ mĕt; ē makēr; ī īce; ĭ bĭt; ō gō; ŏ hŏt; ô ôrb; o͞o mo͞on; ū, ūnite; ŭ ŭp; û ûrn.

Appendix 2
PHYSICAL AND ASTRONOMICAL CONSTANTS

Velocity of light	c = 299,792,458 meters per second
Gravitation constant	G = 6.673 x 10^{-11} $Nm^2 kg^{-2}$
Stefan–Boltzmann constant	σ = 5.67 x 10^{-8} $Wm^{-2} K^{-4}$
Mass of electron	m_e = 9.1094 x 10^{-31} kg
Mass of hydrogen atom	m_H = 1.67352 x 10^{-24} gram
Mass of proton	m_p = 1.67262 x 10^{-27} kg
Astronomical unit	AU = 1.49597870 x 10^{11} m
Parsec	pc = 3.085678 x 10^{16} m
	= 3.261631 light-years
Light-year	LY = 9.460536 x 10^{15} m
Mass of Sun	$M_\odot$ = 1.9891 x 10^{30} kg
Radius of Sun	$**R_\odot$ = 696,265 km
Solar radiation	$L_\odot$ = 3.85 x 10^{26} W
Mass of Earth	$***M_\oplus$ = 5.974 x 10^{24} kg
Equatorial radius of Earth	$R_\oplus$ = 6,378.140 km
Direction of Galactic center	RA = $17^h 45.7^m$, Dec –29°00 ′(2000)
Ephemeris day	d_E = 86,400 seconds
	= 365.2422 ephemeris days
Tropical year (equinox to equinox)	
Sidereal year	= 365.2564 ephemeris days

** $\odot$ = of the Sun
*** $\oplus$ = of the Earth

▶physics.nist.gov/constants◀

MEASUREMENTS AND SYMBOLS

METRIC AND U.S. MEASURES

Approximate Conversions to U.S. Units

Symbol	When You Know	Multiply By	To Find	Symbol
Length				
cm	centimeters	0.39	inches	in
cm	centimeters	0.03	feet	ft
m	meters	1.1	yards	yd
km	kilometers	0.6	miles	ml
Area				
cm^2	square centimeter	0.16	square inches	in^2
m^2	square meters	11	square feet	ft^2
m^2	square meters	1.2	square yards	yd^2
km^2	square kilometer	0.4	square miles	ml^2
Mass (weight)				
g	grams	0.03	ounces	oz
kg	kilograms	2.2	pounds	lb
t	metric ton	1.1	short tons (2000 lb)	
Volume				
L	liters	0.26	gallons	gal
m^3	cubic meters	35	cubic feet	ft^3
m^3	cubic meters	1.3	cubic yards	yd^3
Temperature (exact)				
°C	degrees Celsius	9/5°C + 32	degrees Fahrenheit	°F
K	degrees Kelvin	9/5 K – 460	degrees Fahrenheit	°F

ANGULAR MEASURE

A circle contains 360 degrees, or 360°.
1° contains 60 minutes of arc, or 60'.
1' contains 60 seconds of arc, or 60".

THE GREEK ALPHABET

Alpha	A	α		Nu	N	ν
Beta	B	β		Xi	Ξ	ξ
Gamma	Γ	γ		Omicron	O	o
Delta	Δ	δ		Pi	Π	π
Epsilon	E	ε		Rho	P	ρ
Zeta	Z	ζ		Sigma	Σ	σ
Eta	H	η		Tau	T	τ
Theta	Θ	θ		Upsilon	°	υ
Iota	I	ι		Phi	Φ	φ, φ
Kappa	K	κ		Chi	X	χ
Lambda	Λ	λ		Psi	Ψ	ψ
Mu	M	μ		Omega	Ω	ω

SUN, MOON, PLANET SYMBOLS

☉	Sun	♂	Mars	●	new moon
☾	Moon	♃	Jupiter	☽	first quarter
☿	Mercury	♄	Saturn	○	full moon
♀	Venus	⛢	Uranus	☾	last quarter
⊕	Earth	Ψ	Neptune		

SIGNS OF THE ZODIAC

♈	Aries	0°		♌	Leo	120°		♐ Sagittarius	240°
♉	Taurus	30°		♍	Virgo	150°		♑ Capricornus	270°
♊	Gemini	60°		♎	Libra	180°		♒ Aquarius	300°
♋	Cancer	90°		♏	Scorpius	210°		♓ Pisces	330°

Appendix 4
PERIODIC TABLE OF THE ELEMENTS

You'll find information about each chemical element listed in the Periodic Table at the National Institute of Science and Technology website. ▶physics. nist.gov/data◀

					18 VIIIA 8A
					2 **He** Helium 4.002602(2)

13 IIIA 3A	14 IVA 4A	15 VA 5A	16 VIA 6A	17 VIIA 7A	
5 **B** Boron [10.806;10.821]	6 **C** Carbon [12.0096;12.0116]	7 **N** Nitrogen [14.00643;14.00728]	8 **O** Oxygen [15.99903;15.99977]	9 **F** Fluorine 18.998403163(6)	10 **Ne** Neon 20.1797(6)

11 IB 1B	12 IIB 2B	13 **Al** Aluminum 26.9815386(8)	14 **Si** Silicon [28.084;28.086]	15 **P** Phosphorus 30.973761998(5)	16 **S** Sulfur [32.059;32.076]	17 **Cl** Chlorine [35.446;35.457]	18 **Ar** Argon 39.948(1)
29 **Cu** Copper 63.546(3)	30 **Zn** Zinc 65.38(2)	31 **Ga** Gallium 69.723(1)	32 **Ge** Germanium 72.630(8)	33 **As** Arsenic 74.921595(6)	34 **Se** Selenium 78.971(8)	35 **Br** Bromine [79.901,79.907]	36 **Kr** Krypton 83.798(2)
47 **Ag** Silver 107.8682(2)	48 **Cd** Cadmium 112.414(4)	49 **In** Indium 114.818(1)	50 **Sn** Tin 118.710(7)	51 **Sb** Antimony 121.760(1)	52 **Te** Tellurium 127.60(3)	53 **I** Iodine 126.90447(3)	54 **Xe** Xenon 131.293(6)
79 **Au** Gold 196.966569(5)	80 **Hg** Mercury 200.592(3)	81 **Tl** Thallium [204.382;204.385]	82 **Pb** Lead 207.2(1)	83 **Bi** Bismuth 208.98040(1)	84 **Po** Polonium <209>	85 **At** Astatine <210>	86 **Rn** Radon <222>
111 **Rg** Roentgenium <280>	112 **Cn** Copernicium <285>	113 **Uut** Ununtrium unknown	114 **Fl** Flerovium <289>	115 **Uup** Ununpentium unknown	116 **Lv** Livermorium <293>	117 **Uus** Ununseptium unknown	118 **Uuo** Ununoctium unknown

65 **Tb** Terbium 158.92535(2)	66 **Dy** Dysprosium 162.500(1)	67 **Ho** Holmium 164.93033(2)	68 **Er** Erbium 167.259(3)	69 **Tm** Thulium 168.93422(2)	70 **Yb** Ytterbium 173.054(5)	71 **Lu** Lutetium 174.9668(1)
97 **Bk** Berkelium <247>	98 **Cf** Californium <251>	99 **Es** Einsteinium <252>	100 **Fm** Fermium <257>	101 **Md** Mendelevium <258>	102 **No** Nobelium <259>	103 **Lr** Lawrencium <262>

Appendix 5
THE NEAREST STARS

Star	Distance ly	RA (2000) Dec h m	RA (2000) Dec ° ′	Apparent Magnitude	Spectral Class	Parallax mas	Absolute Magnitude
Sun		— —	— —	−26.72	G		4.9
Proxima	4.2	14 30	−62 41	11.01	M	772	15.5
α Centauri A	4.4	14 40	−60 50	−0.01	G	742	4.3
B	4.4			1.35	K	742	5.7
Barnard's Star	5.9	17 58	+04 42	9.54	M	549	13.2
Wolf 359	7.8	10 56	+07 01	13.46	M	419	16.6
BD +36°2147	8.3	11 03	+35 58	7.49	M	392	10.5
Sirius A	8.6	06 45	−16 43	−1.44	A	379	1.5
B				8.44	DA		11.3
Luyten 726-8A	8.7	01 39	−17 57	12.56	M	373	15.4
B				12.96	M		15.8
Ross 154	9.7	18 50	−23 50	10.70	M	337	13.3
Ross 248	10.3	23 42	+44 09	12.27	M	316	14.8
ε Eridani	10.5	03 33	−09 27	3.72	K	311	6.2
CD-36°15693	10.7	23 06	−35 51	7.35	M	304	9.8
Ross 128	10.9	11 48	+0 48	11.12	M	300	13.5
Luyten 789-6ABC	11.2	22 39	−15 18	12.32	M	290	14.6
61 Cygni A	11.4	21 07	+38 45	5.20	K	287	7.5
Procyon A	11.4	07 39	+05 13	0.40	F	286	2.7
B				10.7	DF		13.0
61 Cygni B	11.4	21 07	+38 45	6.05	K	285	8.3
BD +59 °1915B	11.5	18 43	+59 38	9.70	M	285	12.0
A 11.6				8.94	M	280	11.2

Notes

1. If a star is a visual binary (e.g., Sirius), the letter A designates the brighter component and B, the second component.

2. Star names: Bright stars may be identified by proper names (Sirius), Bayer Greek letters used with constellation names usually in order of brightness (α Centauri), and Flamsteed numbers used with constellation names in order of RA (61 Cygni). Fainter stars are designated by numbers in a catalog such as the Bonner Durchmusterung-BD, Cordoba Durchmusterung-CD, Luyten, Ross and Wolf catalogs. Each catalog has its own numbering system.

3. If a star is a white dwarf (e.g., Sirius B), the letter D precedes its spectral class designation.

Abbreviations

RA = right ascension   h = hours   m = minutes of time   Dec = Declination   ° = degrees
′ = minutes of arc   mas = milliarcseconds = 0″.001   ″ = seconds of arc   ly = light-year

Appendix 6
THE MESSIER OBJECTS

Messier No. (M)	NGC	RA	Dec	Constellation	Description
		h m	° ′		
1	1952	5 34.5	+22 01	Tau	Supernova remnant
2	7089	21 33.5	−0 49	Aqr	Globular cluster
3	5272	13 42.2	+28 23	CVn	Globular cluster
4	6121	16 23.6	−26 32	Sco	Globular cluster
5	5904	15 18.6	+2 05	Ser	Globular cluster
6	6405	17 40.1	−32 13	Sco	Open cluster
7	6475	17 53.9	−34 49	Sco	Open cluster
8	6523	18 03.8	−24 23	Sgr	Diffuse nebula
9	6333	17 19.2	−18 31	Oph	Globular cluster
10	6254	16 57.1	−4 06	Oph	Globular cluster
11	6705	18 51.1	−6 16	Sct	Open cluster
12	6218	16 47.2	−1 57	Oph	Globular cluster
13	6205	16 41.7	+36 28	Her	Globular cluster
14	6402	17 37.6	−3 15	Oph	Globular cluster
15	7078	21 30.0	+12 10	Peg	Globular cluster
16	6611	18 18.8	−13 47	Ser	Open cluster
17	6618	18 20.8	−16 11	Sgr	Diffuse nebula
18	6613	18 19.9	−17 08	Sgr	Open cluster
19	6273	17 02.6	−26 16	Oph	Globular cluster
20	6514	18 02.6	−23 02	Sgr	Diffuse nebula
21	6531	18 04.6	−22 30	Sgr	Open cluster
22	6656	18 36.4	−23 54	Sgr	Globular cluster
23	6494	17 56.8	−19 01	Sgr	Open cluster
24		18 16.9	−18 29	Sgr	*See notes*
25	IC4725	18 31.6	−19 15	Sgr	Open cluster
26	6694	18 45.2	−9 24	Sct	Open cluster
27	6853	19 59.6	+22 43	Vul	Planetary nebula
28	6626	18 24.5	−24 52	Sgr	Globular cluster
29	6913	20 23.9	+38 32	Cyg	Open cluster
30	7099	21 40.4	−23 11	Cap	Globular cluster
31	224	0 42.7	+41 16	And	Spiral galaxy
32	221	0 42.7	+40 52	And	Elliptical galaxy
33	598	1 33.9	+30 39	Tri	Spiral galaxy
34	1039	2 42.0	+42 47	Per	Open cluster
35	2168	6 08.9	+24 20	Gem	Open cluster
36	1960	5 36.1	+34 08	Aur	Open cluster
37	2099	5 52.4	+32 33	Aur	Open cluster
38	1912	5 28.7	+35 50	Aur	Open cluster

(continued)

Messier No. (M)	NGC	RA	Dec	Constellation	Description
		h m	° '		
39	7092	21 32.2	+48 26	Cyg	Open cluster
40		12 22.4	+58 05	UMa	*See notes*
41	2287	6 47.0	−20 44	CMa	Open cluster
42	1976	5 35.4	−5 27	Ori	Diffuse nebula
43	1982	5 35.6	−5 16	Ori	Diffuse nebula
44	2632	8 40.1	+19 59	Cnc	Open cluster
45		3 47.0	+24 07	Tau	Open cluster
46	2437	7 41.8	−14 49	Pup	Open cluster
47	2422	7 36.6	−14 30	Pup	Open cluster
48	2548	8 13.8	−5 48	Hya	Open cluster
49	4472	12 29.8	+8 00	Vir	Elliptical galaxy
50	2323	7 03.2	−8 20	Mon	Open cluster
51	5194-5	13 29.9	+47 12	CVn	Spiral galaxy
52	7654	23 24.2	+61 35	Cas	Open cluster
53	5024	13 12.9	+18 10	Com	Globular cluster
54	6715	18 55.1	−30 29	Sgr	Globular cluster
55	6809	19 40.0	−30 58	Sgr	Globular cluster
56	6779	19 16.6	+30 11	Lyr	Globular cluster
57	6720	18 53.6	+33 02	Lyr	Planetary nebula
58	4579	12 37.7	+11 49	Vir	Spiral galaxy
59	4621	12 42.0	+11 39	Vir	Elliptical galaxy
60	4649	12 43.7	+11 33	Vir	Elliptical galaxy
61	4303	12 21.9	+4 28	Vir	Spiral galaxy
62	6266	17 01.2	−30 07	Oph	Globular cluster
63	5055	13 15.8	+42 02	CVn	Spiral galaxy
64	4826	12 56.7	+21 41	Com	Spiral galaxy
65	3623	11 18.9	+13 05	Leo	Spiral galaxy
66	3627	11 20.2	+12 59	Leo	Spiral galaxy
67	2682	8 50.4	+11 49	Cnc	Open cluster
68	4590	12 39.5	−26 45	Hya	Globular cluster
69	6637	18 31.4	−32 21	Sgr	Globular cluster
70	6681	18 43.2	−32 18	Sgr	Globular cluster
71	6838	19 53.8	+18 47	Sge	Globular cluster
72	6981	20 53.5	−12 32	Aqr	Globular cluster
73	6994	20 58.9	−12 38	Aqr	*See notes*
74	628	1 36.7	+15 47	Psc	Spiral galaxy
75	6864	20 06.1	−21 55	Sgr	Globular cluster
76	650-1	1 42.4	+51 34	Per	Planetary nebula
77	1068	2 42.7	−0 01	Cet	Spiral galaxy
78	2068	5 46.7	+0 03	Ori	Diffuse nebula
79	1904	5 24.5	−24 33	Lep	Globular cluster
80	6093	16 17.0	−22 59	Sco	Globular cluster
81	3031	9 55.6	+69 04	UMa	Spiral galaxy
82	3034	9 55.8	+69 41	UMa	Irregular galaxy
83	5236	13 37.0	−29 52	Hya	Spiral galaxy
84	4374	12 25.1	+12 53	Vir	Elliptical galaxy
85	4382	12 25.4	+18 11	Com	Elliptical galaxy
86	4406	12 26.2	+12 57	Vir	Elliptical galaxy
87	4486	12 30.8	+12 24	Vir	Elliptical galaxy

Messier No. (M)	NGC	RA	Dec	Constellation	Description
		h m	° ′		
88	4501	12 32.0	+14 25	Com	Spiral galaxy
89	4552	12 35.7	+12 33	Vir	Elliptical galaxy
90	4569	12 36.8	+13 10	Vir	Spiral galaxy
91	4548	12 35.4	+14 30	Com	Spiral galaxy
92	6341	17 17.1	+43 08	Her	Globular cluster
93	2447	7 44.6	−23 52	Pup	Open cluster
94	4736	12 50.9	+41 07	CVn	Spiral galaxy
95	3351	10 44.0	+11 42	Leo	Spiral galaxy
96	3368	10 46.8	+11 49	Leo	Spiral galaxy
97	3587	11 14.8	+55 01	UMa	Planetary nebula
98	4192	12 13.8	+14 54	Com	Spiral galaxy
99	4254	12 18.8	+14 25	Com	Spiral galaxy
100	4321	12 22.9	+15 49	Com	Spiral galaxy
101	5457	14 03.2	+54 21	UMa	Spiral galaxy
102					*See notes*
103	581	1 33.2	+60 42	Cas	Open cluster
104	4594	12 40.0	−11 37	Vir	Spiral galaxy
105	3379	10 47.8	+12 35	Leo	Elliptical galaxy
106	4258	12 19.0	+47 18	CVn	Spiral galaxy
107	6171	16 32.5	−13 03	Oph	Globular cluster
108	3556	11 11.5	+55 40	UMa	Spiral galaxy
109	3992	11 57.6	+53 23	UMa	Spiral galaxy
110	205	0 40.4	+41 41	And	Elliptical galaxy

FAVORITES

M1	Crab Nebula
M8	Lagoon Nebula; contains a star cluster
M11	Wild Duck Cluster
M16	Surrounded by the Eagle Nebula
M17	Omega Nebula
M20	Trifid Nebula
M24	Star field in Sagittarius, containing the open cluster NGC 6603
M27	Dumbbell Nebula
M31	Andromeda Galaxy
M40	Faint double star Winnecke 4, mags. 9.0 and 9.6
M42, M43	Orion Nebula
M44	Praesepe, the Beehive Cluster
M45	The Pleiades; no NGC or IC number
M51	Whirlpool Galaxy
M57	Ring Nebula
M64	Black Eye Galaxy
M73	Small group of four faint stars
M97	Owl Nebula
M102	Duplicate of M101
M104	Sombrero Galaxy

Source: A. Hirshfeld and R. W. Sinnott (eds.), *Sky Catalogue 2000.0*, Vol. 2 (Sky Publishing Corp./Cambridge University Press, 1985).

INDEX

STAR AND MOON MAPS

SPRING SKIES

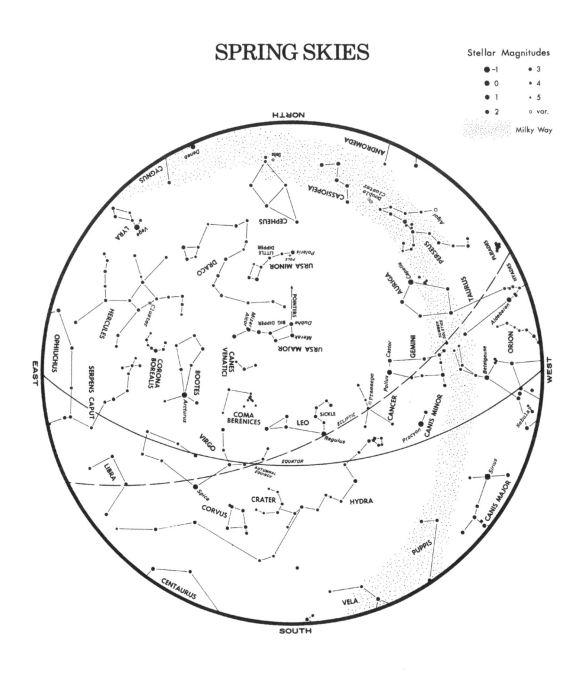

Stellar Magnitudes

● -1 ● 3
● 0   ● 4
● 1   · 5
● 2   ○ var.

▓ Milky Way

Chart by George Lovi

TIME SCHEDULE

Late March	11 p.m.	Early May	8 p.m.
Early April	10 p.m.	Late May	7 p.m.
Late April	9 p.m.	Early June	6 p.m.

STANDARD TIME

SUMMER SKIES

Stellar Magnitudes

- ● −1
- ● 0
- ● 1
- ● 2
- ● 3
- ● 4
- · 5
- ○ var.

Milky Way

Chart by George Lovi

TIME SCHEDULE

Late June	11 p.m.	Early August	8 p.m.
Early July	10 p.m.	Late August	7 p.m.
Late July	9 p.m.	Early September	6 p.m.

STANDARD TIME

AUTUMN SKIES

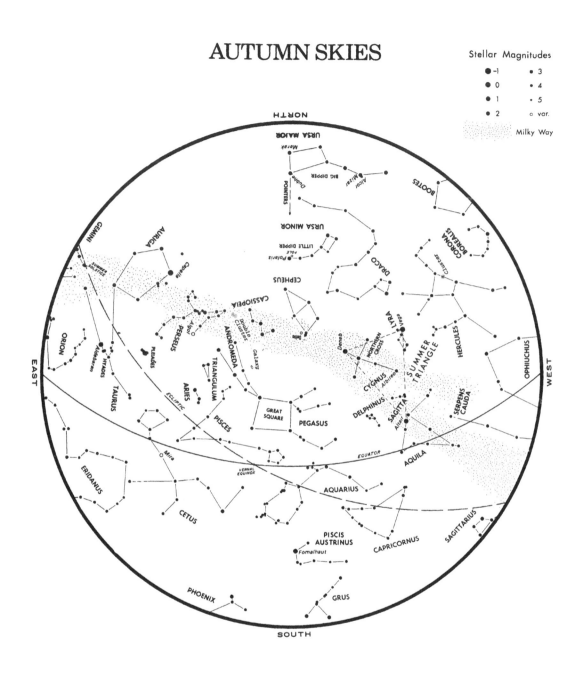

Stellar Magnitudes

- −1
- 0
- 1
- 2
- 3
- 4
- 5
- var.

Milky Way

Chart by George Lovi

TIME SCHEDULE

Late September	11 p.m.	Early November	8 p.m.
Early October	10 p.m.	Late November	7 p.m.
Late October	9 p.m.	Early December	6 p.m.

STANDARD TIME

WINTER SKIES

Stellar Magnitudes

- ● -1
- ● 0
- ● 1
- ● 2
- ● 3
- ● 4
- · 5
- ○ var.

Milky Way

Chart by George Lovi

TIME SCHEDULE

Late December	11 p.m.	Early February	8 p.m.
Early January	10 p.m.	Late February	7 p.m.
Late January	9 p.m.	Early March	6 p.m.

STANDARD TIME

MOON MAP

NORTH

J. Herschel

Plato

MARE FRIGORIS

Aristoteles

Atlas

Alps

Eudoxus

Aristillus

Posidonius

Cleomedes

Archimedes

Autolycus

MARE
SERENITATIS

MARE
CRISIUM

Struve

MARE IMBRIUM

Aristarchus

EAST

Apennines

Plinius

Carpathians

Eratosthenes

Manilius

MARE
TRANQUILLITATIS

Kepler

Copernicus

Delambre

MARE
FECUNDITATIS

WEST

OCEANUS
PROCELLARUM

Riphaeus
Mts.

Ptolemaeus

Albategnius

Theophilus

Longrenus

Grimaldi

Alphonsus

Cyrillus

MARE
NECTARIS

Gassendi

Bullialdus

Arzachel

Catharina

Mersenius

MARE
HUMORUM

MARE
NUBIUM

Straight
Wall

Sacrobosco

Piccolomini

Petavius

Vieta

Walter

Tycho

Janssen

Schickard

Stöfler

Maurolycus

Longomontanus

Vlacq

Schiller

Clavius

Manzinus

SOUTH

KEY

Prominent crater (with
or without interior peak)

Prominent mountain
range

Chart by George Lovi

Printed in the USA
CPSIA information can be obtained
at www.ICGtesting.com
JSHW071322221024
72173JS00010B/1617